THE NUMBERS BEHIND NUMB3RS

DR. KEITH DEVLIN is executive director of Stanford University's Center for the Study of Language and Information and a consulting professor of mathematics at Stanford. Devlin has a B.Sc. degree in Mathematics from King's College London (1968) and a Ph.D. in Mathematics from the University of Bristol (1971). He is a fellow of the American Association for the Advancement of Science, a World Economic Forum fellow, and a former member of the Mathematical Sciences Education Board of the U.S. National Academy of Sciences. The author of twenty-five books, Devlin has been a regular contributor to National Public Radio's popular program *Weekend Edition*, where he is known as "the Math Guy" in his on-air conversations with host Scott Simon. His monthly column, "Devlin's Angle," appears on Mathematical Association of America's web journal *MAA Online*.

DR. GARY LORDEN is a professor in the mathematics department of the California Institute of Technology in Pasadena. He graduated from Caltech with a B.S. in mathematics in 1962, received his Ph.D. in mathematics from Cornell University in 1966, and taught at Northwestern University before returning to Caltech in 1968. A fellow of the Institute of Mathematical Statistics, Lorden has taught statistics, probability, and other mathematics at all levels from freshman to doctoral. Lorden has also been active as a consultant and expert witness in mathematics and statistics for government agencies and laboratories, private companies, and law firms. For many years he consulted for Caltech's Jet Propulsion Laboratory for their space exploration programs. He has participated in highly classified research projects aimed at enhancing the ability of government agencies (such as the NSA) to protect national security. Lorden is the chief mathematics consultant for the CBS TV series *NUMB3RS*.

THE NUMBERS BEHIND *NUMB3RS*

Solving Crime with Mathematics

Keith Devlin, Ph.D.

and

Gary Lorden, Ph.D.

A PLUME BOOK

PLUME
Published by Penguin Group
Penguin Group (USA) Inc., 375 Hudson Street, New York, New York 10014, U.S.A. • Penguin Group (Canada), 90 Eglinton Avenue East, Suite 700, Toronto, Ontario, Canada M4P 2Y3 (a division of Pearson Penguin Canada Inc.) • Penguin Books Ltd., 80 Strand, London WC2R 0RL, England • Penguin Ireland, 25 St. Stephen's Green, Dublin 2, Ireland (a division of Penguin Books Ltd.) • Penguin Group (Australia), 250 Camberwell Road, Camberwell, Victoria 3124, Australia (a division of Pearson Australia Group Pty. Ltd.) • Penguin Books India Pvt. Ltd., 11 Community Centre, Panchsheel Park, New Delhi – 110 017, India • Penguin Books (NZ), 67 Apollo Drive, Rosedale, North Shore 0745, Auckland, New Zealand (a division of Pearson New Zealand Ltd.) • Penguin Books (South Africa) (Pty.) Ltd., 24 Sturdee Avenue, Rosebank, Johannesburg 2196, South Africa

Penguin Books Ltd., Registered Offices: 80 Strand, London WC2R 0RL, England

First published by Plume, a member of Penguin Group (USA) Inc.

First Printing, September 2007
10 9 8 7 6 5 4 3 2

Illustration credits appear on page 244.

LIBRARY OF CONGRESS CATALOGING-IN-PUBLICATION DATA

Devlin, Keith J.
The numbers behind NUMB3RS: solving crime with mathematics / Keith Devlin, Gary Lorden.
p. cm.
ISBN 978-0-452-28857-7
1. Criminal investigation. 2. Mathematical statistics. 3. Criminal investigation—Data processing. I. Title: Numbers behind numbers. II. Lorden, Gary. III. Title.

HV8073.5.D485 2007
363.2501'5195—dc22
2007018115

Printed in the United States of America
Set in Dante MT
Designed by Joseph Rutt

Acknowledgments

The authors want to thank *NUMB3RS* creators Cheryl Heuton and Nick Falacci for creating Charlie Eppes, television's first mathematics superhero, and succeeding brilliantly in putting math on television in prime time. Their efforts have been joined by a stellar team of other writers, actors, producers, directors, and specialists whose work has inspired us to write this book. The gifted actor David Krumholtz has earned the undying love of mathematicians everywhere for bringing Charlie to life in a way that has led millions of people to see mathematics in a completely new light. Thanks also to *NUMB3RS* researchers Andy Black and Matt Kolokoff for being wonderful to work with in coming up with endless applications of mathematics to make the writers' dreams come true.

We wish to express our particular thanks to mathematician Dr. Lenny Rudin of Cognitech, one of the world's foremost experts on image enhancement, for considerable help with Chapter 5 and for providing the images we show in that chapter.

Finally, Ted Weinstein, our agent, found us an excellent publisher in David Cashion of Plume, and both worked tirelessly to turn a manuscript that we felt was as reader-friendly as possible, given that this is a math book, into one that, we have to acknowledge, is now a lot more so!

Keith Devlin, Palo Alto, CA
Gary Lorden, Pasadena, CA

Contents

INTRODUCTION

The Hero Is a *Mathematician?*

On January 23, 2005, a new television crime series called *NUMB3RS* debuted. Created by the husband-and-wife team Nick Falacci and Cheryl Heuton, the series was produced by Paramount Network Television and acclaimed Hollywood veterans Ridley and Tony Scott, whose movie credits include *Alien*, *Top Gun*, and *Gladiator*. Throughout its run, *NUMB3RS* has regularly beat out the competition to be the most watched series in its time slot on Friday nights.

What has surprised many is that one of the show's two heroes is a mathematician, and much of the action revolves around mathematics, as professor Charlie Eppes uses his powerful skills to help his older brother, Don, an FBI agent, identify and catch criminals. Many viewers, and several critics, have commented that the stories are entertaining, but the basic premise is far-fetched: You simply can't use math to solve crimes, they say. As this book proves, they are wrong. You can use math to solve crimes, and law enforcement agencies do—not in every instance to be sure, but often enough to make math a powerful weapon in the never-ending fight against crime. In fact, the very first episode of the series was closely based on a real-life case, as we will discuss in the next chapter.

Our book sets out to describe, in a nontechnical fashion, some of the major mathematical techniques currently available to the police, CIA, and FBI. Most of these methods have been mentioned during episodes of *NUMB3RS*, and while we frequently link our explanations to what was depicted on the air, our focus is on the mathematical techniques and how they can be used in law enforcement. In addition we describe some real-life cases where mathematics played a role in solving a crime that have not been used in the TV series—at least not directly.

In many ways, *NUMB3RS* is similar to good science fiction, which is based on correct physics or chemistry. Each week, *NUMB3RS* presents a dramatic story in which realistic mathematics plays a key role in the narrative. The producers of *NUMB3RS* go to great lengths to ensure that the mathematics used in the scripts is correct and that the applications shown are possible. Although some of the cases viewers see are fictional, they certainly could have happened, and in some cases very well may. Though the TV series takes some dramatic license, this book does not. In *The Numbers Behind NUMB3RS*, you will discover the mathematics that can be, and is, used in fighting real crime and catching actual criminals.

THE NUMBERS BEHIND *NUMB3RS*

CHAPTER

1 Finding the Hot Zone
Criminal Geographic Profiling

FBI Special Agent Don Eppes looks again at the large street map of Los Angeles spread across the dining-room table of his father's house. The crosses inked on the map show the locations where, over a period of several months, a brutal serial killer has struck, raping and then murdering a number of young women. Don's job is to catch the killer before he strikes again. But the investigation has stalled. Don is out of clues, and doesn't know what to do next.

"Can I help?" The voice is that of Don's younger brother, Charlie, a brilliant young professor of mathematics at the nearby university CalSci. Don has always been in awe of his brother's incredible ability at math, and frankly would welcome any help he can get. But . . . help from a mathematician?

"This case isn't about numbers, Charlie." The edge in Don's voice is caused more by frustration than anger, but Charlie seems not to notice, and his reply is totally matter-of-fact but insistent: "*Everything* is numbers."

Don is not convinced. Sure, he has often heard Charlie say that mathematics is all about patterns—identifying them, analyzing them, making predictions about them. But it didn't take a math genius to see that the crosses on the map were scattered haphazardly. There was no pattern, no way anyone could predict where the next cross would go—the exact location where the next young girl would be attacked. Maybe it would occur that very evening. If only there were some regularity to the arrangement of the crosses, a pattern that could be captured with a mathematical equation, the way Don remembers from his schooldays that the equation $x^2 + y^2 = 9$ describes a circle.

Looking at the map, even Charlie has to agree there is no way to use math to predict where the killer would strike next. He strolls over to the window and stares out across the garden, the silence of the evening broken only by the continual *flick-flick-flick-flick* of the automatic sprinkler watering the lawn. Charlie's eyes see the sprinkler but his mind is far away. He had to admit that Don was probably right. Mathematics could be used to do lots of things, far more than most people realized. But in order to use math, there had to be some sort of pattern.

Flick-flick-flick-flick. The sprinkler continued to do its job. There was the brilliant mathematician in New York who used mathematics to study the way the heart works, helping doctors spot tiny irregularities in a heartbeat before the person has a heart attack.

Flick-flick-flick-flick. There were all those mathematics-based computer programs the banks utilized to track credit card purchases, looking for a sudden change in the pattern that might indicate identity theft or a stolen card.

Flick-flick-flick-flick. Without clever mathematical algorithms, the cell phone in Charlie's pocket would have been twice as big and a lot heavier.

Flick-flick-flick-flick. In fact, there was scarcely any area of modern life that did not depend, often in a crucial way, on mathematics. But there had to be a pattern, otherwise the math can't get started.

Flick-flick-flick-flick. For the first time, Charlie notices the sprinkler, and suddenly he knows what to do. He has his answer. He could help solve Don's case, and the solution has been staring him in the face all along. He just had not realized it.

He drags Don over to the window. "We've been asking the wrong question," he says. "From what you know, there's no way you can predict where the killer will strike next." He points to the sprinkler. "Just like, no matter how much you study where each drop of water hits the grass, there's no way you can predict where the next drop will land. There's too much uncertainty." He glances at Don to make sure his older brother is listening. "But suppose you could not see the sprinkler, and all you had to go on was the pattern of where all the drops landed. Then, using math, you could work out exactly where the sprinkler must be. You can't use the pattern of drops to predict forward to the next drop, but you can use it to work backward to the source. It's the same with your killer."

Don finds it difficult to accept what his brother seems to be suggesting. "Charlie, are you telling me you can figure out where the killer lives?"

Charlie's answer is simple: "Yes."

Don is still skeptical that Charlie's idea can really work, but he's impressed by his brother's confidence and passion, and so he agrees to let him assist with the investigation.

Charlie's first step is to learn some basic facts from the science of criminology: First, how do serial killers behave? Here, his years of experience as a mathematician have taught him how to recognize the key factors and ignore all the others, so that a seemingly complex problem can be reduced to one with just a few key variables. Talking with Don and the other agents at the FBI office where his elder brother works, he learns, for instance, that violent serial criminals exhibit certain tendencies in selecting locations. They tend to strike close to their home, but not too close; they always set a "buffer zone" around their residence where they will not strike, an area that is too close for comfort; outside that comfort zone, the frequency of crime locations decreases as the distance from home increases.

Then, back in his office in the CalSci mathematics department, Charlie gets to work in earnest, feverishly covering his blackboards with mathematical equations and formulas. His goal: to find the mathematical key to determine a "hot zone"—an area on the map, derived from the crime locations, where the perpetrator is most likely to live.

As always when he works on a difficult mathematical problem, the hours fly by as Charlie tries out many unsuccessful approaches. Then, finally, he has an idea he thinks should work. He erases his previous chalk scribbles one more time and writes this complicated-looking formula on the board:*

$$p_{ij} = k\sum_{n=1}^{c}\left[\frac{\phi}{\left(|x_i - x_n| + |y_j - y_n|\right)^f} + \frac{(1-\phi)\left(B^{g-f}\right)}{\left(2B - |x_i - x_n| - |y_j - y_n|\right)^g}\right]$$

*We'll take a closer look at this formula in a moment.

"That should do the trick," he says to himself.

The next step is to fine-tune his formula by checking it against examples of past serial crimes Don provides him with. When he inputs the crime locations from those previous cases into his formula, does it accurately predict where the criminals lived? This is the moment of truth, when Charlie will discover whether his mathematics reflects reality. Sometimes it doesn't, and he learns that when he first decided which factors to take into account and which to ignore, he must have got it wrong. But this time, after Charlie makes a few minor adjustments, the formula seems to work.

The next day, bursting with energy and conviction, Charlie shows up at the FBI offices with a printout of the crime-location map with the "hot zone" prominently displayed. Just as the equation $x^2 + y^2 = 9$ that Don remembered from his schooldays describes a circle, so that when the equation is fed into a suitably programmed computer it will draw the circle, so too when Charlie fed his new equation into his computer, it also produced a picture. Not a circle this time—Charlie's equation is much more complicated. What it gave was a series of concentric colored regions drawn on Don's crime map of Los Angeles, regions that homed in on the hot zone where the killer lives.

Having this map will still leave a lot of work for Don and his colleagues, but finding the killer is no longer like looking for a needle in a haystack. Thanks to Charlie's mathematics, the haystack has suddenly dwindled to a mere sackful of hay.

Charlie explains to Don and the other FBI agents working the case that the serial criminal has tried not to reveal where he lives, picking victims in what he thinks is a random pattern of locations, but that the mathematical formula nevertheless reveals the truth: a hot zone in which the criminal's residence is located, to a very high probability. Don and the team decide to investigate men within a certain range of ages, who live in the hot zone, and use surveillance and stealth tactics to obtain DNA evidence from the suspects' discarded cigarette butts, drinking straws, and the like, which can be matched with DNA from the crime-scene investigations.

Within a few days—and a few heart-stopping moments—they have their man. The case is solved. Don tells his younger brother, "That's *some formula* you've got there, Charlie."

FACT OR FICTION?

Leaving out a few dramatic twists, the above is what the TV audience saw in the very first episode of *NUMB3RS*, broadcast on January 23, 2005. Many viewers could not believe that mathematics could help capture a criminal in this way. In fact, that entire first episode was based fairly closely on a real case in which a single mathematical equation was used to identify the hot zone where a criminal lived. It was the very equation, reproduced above, that viewers saw Charlie write on his blackboard.

The real-life mathematician who produced that formula is named Kim Rossmo. The technique of using mathematics to predict where a serial criminal lives, which Rossmo helped to establish, is called geographic profiling.

In the 1980s Rossmo was a young constable on the police force in Vancouver, Canada. What made him unusual for a police officer was his talent for mathematics. Throughout school he had been a "math whiz," the kind of student who makes fellow students, and often teachers, a little nervous. The story is told that early in the twelfth grade, bored with the slow pace of his mathematics course, he asked to take the final exam in the second week of the semester. After scoring one hundred percent, he was excused from the remainder of the course.

Similarly bored with the typical slow progress of police investigations involving violent serial criminals, Rossmo decided to go back to school, ending up with a Ph.D. in criminology from Simon Fraser University, the first cop in Canada to get one. His thesis advisers, Paul and Patricia Brantingham, were pioneers in the development of mathematical models (essentially sets of equations that describe a situation) of criminal behavior, particularly those that describe where crimes are most likely to occur based on where a criminal lives, works, and plays. (It was the Brantinghams who noticed the location patterns of serial criminals that TV veiwers saw Charlie learning about from Don and his FBI colleagues.)

Rossmo's interest was a little different from the Brantinghams'. He did not want to study patterns of criminal behavior. As a police officer, he wanted to use actual data about the locations of crimes linked to a single unknown perpetrator as an *investigative tool* to help the police find the criminal.

Rossmo had some initial successes in re-analyzing old cases, and after receiving his Ph.D. and being promoted to detective inspector, he pursued his interest in developing better mathematical methods to do what he came to call criminal geographic targeting (CGT). Others called the method "geographic profiling," since it complemented the well-known technique of "psychological profiling" used by investigators to find criminals based on their motivations and psychological characteristics. Geographic profiling attempts to locate a likely base of operation for a criminal by analyzing the locations of their crimes.

Rossmo hit upon the key idea behind his seemingly magic formula while riding on a bullet train in Japan one day in 1991. Finding himself without a notepad to write on, he scribbled it on a napkin. With later refinements, the formula became the principal element of a computer program Rossmo wrote, called Rigel (pronounced RYE-gel, and named after the star in the constellation Orion, the Hunter). Today, Rossmo sells Rigel, along with training and consultancy, to police and other investigative agencies around the world to help them find criminals.

When Rossmo describes how Rigel works to a law enforcement agency interested in the program, he offers his favorite metaphor—that of determining the location of a rotating lawn sprinkler by analyzing the pattern of the water drops it sprays on the ground. When *NUMB3RS* cocreators Cheryl Heuton and Nick Falacci were working on their pilot episode, they took Rossmo's own metaphor as the way Charlie would hit upon the formula and explain the idea to his brother.

Rossmo had some early successes dealing with serial crime investigations in Canada, but what really made him a household name among law enforcement agencies all over North America was the case of the South Side Rapist in Lafayette, Louisiana.

For more than ten years, an unknown assailant, his face wrapped bandit-style in a scarf, had been stalking women in the town and assaulting them. In 1998 the local police, snowed under by thousands of tips and a corresponding number of suspects, brought Rossmo in to help. Using Rigel, Rossmo analyzed the crime-location data and produced a map much like the one Charlie displayed in *NUMB3RS*, with bands of color indicating the hot zone and its increasingly hot interior rings. The map enabled police to narrow down the hunt to half a square mile and about a dozen suspects. Undercover officers combed the hot zone using the same techniques portrayed in *NUMB3RS*, to obtain DNA samples of all males of the right age range in the area.

Frustration set in when each of the suspects in the hot zone was cleared by DNA evidence. But then they got lucky. The lead investigator, McCullan "Mac" Gallien, received an anonymous tip pointing to a very unlikely suspect—a sheriff's deputy from a nearby department. As just one more tip on top of the mountain he already had, Mac was inclined to just file it, but on a whim he decided to check the deputy's address. Not even close to the hot zone. Still something niggled him, and he dug a little deeper. And then he hit the jackpot. The deputy had previously lived at another address—right in the hot zone! DNA evidence was collected from a cigarette butt, and it matched that taken from the crime scenes. The deputy was arrested, and Rossmo became an instant celebrity in the crime-fighting world.

Interestingly, when Heuton and Falacci were writing the pilot episode of *NUMB3RS*, based on this real-life case, they could not resist incorporating the same dramatic twist at the end. When Charlie first applies his formula, no DNA matches are found among the suspects in the hot zone, as happened with Rossmo's formula in Lafayette. Charlie's belief in his mathematical analysis is so strong that when Don tells him the search has drawn a blank, he initially refuses to accept this outcome. "You must have missed him," he says.

Frustrated and upset, Charlie huddles with Don at their father Alan's house, and Alan says, "I know the problem can't be the math, Charlie. It must be something else." This remark spurs Don to realize that finding the killer's *residence* may be the wrong goal. "If you tried to find me where I *live*, you would probably fail because I'm almost never there," he notes. "I'm usually at *work*." Charlie seizes on this notion to pursue a different line of attack, modifying his calculations to look for *two* hot zones, one that might contain the killer's residence and the other his place of work. This time Charlie's math works. Don manages to identify and catch the criminal just before he kills another victim.

These days, Rossmo's company ECRI (Environmental Criminology Research, Inc.) offers the patented computer package Rigel along with training in how to use it effectively to solve crimes. Rossmo himself travels around the world, to Asia, Africa, Europe, and the Middle East, assisting in criminal investigations and giving lectures to police and criminologists. Two years of training, by Rossmo or one of his assistants, is required to learn to adapt the use of the program to the idiosyncrasies of a particular criminal's behavior.

Rigel does not score a big win every time. For example, Rossmo was called in on the notorious Beltway Sniper case when, during a three-week period in October 2002, ten people were killed and three others critically injured by what turned out to be a pair of serial killers operating in and around the Washington, D.C., area. Rossmo concluded that the sniper's base was somewhere in the suburbs to the north of Washington, but it turned out that the two killers did not live in the area and moved too often to be located by geographic profiling.

The fact that Rigel does not always work will not come as a surprise to anyone familiar with what happens when you try to apply mathematics to the messy real world of people. Many people come away from their high school experience with mathematics thinking that there is a right way and a wrong way to use math to solve a problem—in too many cases with the teacher's way being the right one and their own attempts being the wrong one. But this is rarely the case. Mathematics will always give you the correct answer (if you do the math right) when you apply it to very well-defined physical situations, such as calculating how much fuel a jet needs to fly from Los Angeles to New York. (That is, the math will give you the right answer provided you start with accurate data about the total weight of the plane, passengers, and cargo, the prevailing winds, and so forth. Missing a key piece of input data to incorporate into the mathematical equations will almost always result in an inaccurate answer.) But when you apply math to a social problem, such as a crime, things are rarely so clear-cut.

Setting up equations that capture elements of some real-life activity is called constructing a "mathematical model." In constructing a *physical* model of something, say an aircraft to study in a wind tunnel, the important thing is to get everything right, apart from the size and the materials used. In constructing a mathematical model, the idea is to get the appropriate *behavior* right. For example, to be useful, a mathematical model of the weather should predict rain for days when it rains and predict sunshine on sunny days. Constructing the model in the first place is usually the hard part. "Doing the math" with the model—i.e., solving the equations that make up the model—is generally much easier, especially when using computers. Mathematical models of the weather often fail because the weather is simply far too complicated (in everyday language, it's "too unpredictable") to be captured by mathematics with great accuracy.

As we shall see in later chapters, there is usually no such thing as "one correct way" to use mathematics to solve problems in the real world, particularly problems involving people. To try to meet the challenges that confront Charlie in *NUMB3RS*—locating criminals, tracing the spread of a disease or of counterfeit money, predicting the target selection of terrorists, and so on—a mathematician cannot merely write down an equation and solve it. There is a considerable art to the process of assembling information and data, selecting mathematical variables that describe a situation, and then modeling it with a set of equations. And once a mathematician has constructed a model, there is still the matter of solving it in some way, by approximations or calculations or computer simulations. Every step in the process requires judgment and creativity. No two mathematicians working independently, however brilliant, are likely to produce identical results, if indeed they can produce useful results at all.

It is not surprising, then, that in the field of geographic profiling, Rossmo has competitors. Dr. Grover M. Godwin of the Justice Center at the University of Alaska, author of the book *Hunting Serial Predators*, has developed a computer package called Predator that uses a branch of mathematical statistics called multivariate analysis to pinpoint a serial killer's home base by analyzing the locations of crimes, where the victims were last seen, and where the bodies were discovered. Ned Levine, a Houston-based urban planner, developed a program called Crimestat for the National Institute of Justice, a research branch of the U.S. Justice Department. It uses something called spatial statistics to analyze serial-crime data, and it can also be applied to help agents understand such things as patterns of auto accidents or disease outbreaks. And David Canter, a professor of psychology at the University of Liverpool in England, and the director of the Centre for Investigative Psychology there, has developed his own computer program, Dragnet, which he has sometimes offered free to researchers. Canter has pointed out that so far no one has performed a head-to-head comparison of the various math/computer systems for locating serial criminals based on applying them in the same cases, and he has claimed in interviews that in the long run, his program and others will prove to be at least as accurate as Rigel.

ROSSMO'S FORMULA

Finally, let's take a closer look at the formulas Rossmo scribbled down on that paper napkin on the bullet train in Japan back in 1991.

$$p_{ij} = k\sum_{n=1}^{c}\left[\frac{\phi}{\left(|x_i - x_n| + |y_j - y_n|\right)^f} + \frac{(1-\phi)\left(B^{g-f}\right)}{\left(2B - |x_i - x_n| - |y_j - y_n|\right)^g}\right],$$

To understand what it means, imagine a grid of little squares superimposed on the map, each square having two numbers that locate it: what row it's in and what column it's in, "*i*" and "*j*". The probability, p_{ij}, that the killer's residence is in that square is written on the left side of the equation, and the right side shows how to calculate it. The crime locations are represented by map coordinates, (x_1,y_1) for the first crime, (x_2,y_2) for the second crime, and so on. What the formula says is this:

To get the probability p_{ij} for the square in row "*i*", column "*j*" of the grid, first calculate how far you have to go to get from the center point (x_i,y_j) of that square to each crime location (x_n,y_n). The little "n" here stands for any one of the crime locations—n=1 means "first crime," n=2 means "second crime," and so on. The answer to the question of how far you have to go is:

$$|x_i - x_n| + |y_j - y_n|$$

and this is used in two ways.

Reading from left to right in the formula, the first way is to put that distance in the denominator, with φ in the numerator. The distance is raised to the power *f*. The choice of what number to use for this *f* will be based on what works best when the formula is checked against data on past crime patterns. (If you take *f* = 2, for example, then that part of the formula will resemble the "inverse square law" that describes the force of gravity.) This part of the formula expresses the idea that the probability of crime locations *decreases* as the distance increases, once outside of the buffer zone.

The second way the formula uses the "traveling distance" of each crime involves the buffer zone. In the second fraction, you *subtract* the distance from 2B, where B is a number that will be chosen to describe the size of the buffer zone, and you use that subtraction result in the second fraction. The subtraction produces *smaller* answers as the distance increases, so that after raising those answers to another power, *g*, in the denominator of the second part of the formula, you get *larger* results.

Together, the first and second parts of the formula perform a sort of "balancing act," expressing the fact that as you move away from the criminal's base, the probability of crimes first *increases* (as you move through the buffer zone) and then *decreases*. The two parts of the formula are combined using a fancy mathematical notation, the Greek letter Σ standing for "sum (add up) the contributions from each of the crimes to the evaluation of the probability for the '*ij*' grid square." The Greek letter φ is used in the two parts as a way of placing more "weight" on one part or the other. A larger choice of φ puts more weight on the phenomenon of "decreasing probability as distance increases," whereas a smaller φ emphasizes the effect of the buffer zone.

Once the formula is used to calculate the probabilities, p_{ij}, of all of the little squares in the grid, it's easy to make a hot zone map. You just color the squares, with the highest probabilities bright yellow, slightly smaller probabilities orange, then red, and so on, leaving the squares with low probability uncolored.

Rossmo's formula is a good example of the art of using mathematics to describe incomplete knowledge of real-world phenomena. Unlike the law of gravity, which through careful measurements can be observed to operate *the same way every time*, descriptions of the behavior of individual human beings are at best approximate and uncertain. When Rossmo checked out his formula on past crimes, he had to find the best fit of his formula to those data by choosing different possible values of *f* and *g*, and of B and φ. He then used those findings in analyzing future crime patterns, still allowing for further fine-tuning in each new investigation.

Rossmo's method is definitely not rocket science—space travel depends crucially on always getting the right answer with great accuracy. But it is nevertheless science. It does not work every time, and the answers it gives are probabilities. But in crime detection and other domains involving human behavior, knowing those probabilities can sometimes make all the difference.

CHAPTER

2 Fighting Crime with Statistics 101

THE ANGEL OF DEATH

By 1996, Kristen Gilbert, a thirty-three-year-old divorced mother of two sons, ages seven and ten, and a nurse in Ward C at the Veteran's Affairs Medical Center in Northampton, Massachusetts, had built up quite a reputation among her colleagues at the hospital. On several occasions she was the first one to notice that a patient was going into cardiac arrest and to sound a "code blue" to bring the emergency resuscitation team. She always stayed calm, and was competent and efficient in administering to the patient. Sometimes she would give the patient an injection of the heart-stimulant drug epinephrine to attempt to restart the heart before the emergency team arrived, occasionally saving the patient's life in this way. The other nurses had given her the nickname "Angel of Death."

But that same year, three nurses approached the authorities to express their growing suspicions that something was not quite right. There had been just too many deaths from cardiac arrest in that particular ward, they felt. There had also been several unexplained shortages of epinephrine. The nurses were starting to fear that Gilbert was giving the patients large doses of the drug to bring on the heart attacks in the first place, so that she could play the heroic role of trying to save them. The "Angel of Death" nickname was beginning to sound more apt than they had first intended.

The hospital launched an investigation, but found nothing untoward. In particular, the number of cardiac deaths at the unit was broadly in line with the rates at other VA hospitals, they said. Despite the findings of the initial investigation, however, the staff at the hospital remained suspicious, and eventually a second investigation was begun. This included bringing in a professional statistician, Stephen Gehlbach of the University of Massachusetts, to take a closer look at the unit's cardiac arrest and mortality figures. Largely as a result of Gehlbach's analysis, in 1998 the U.S. Attorney's Office decided to convene a grand jury to hear the evidence against Gilbert.

Part of the evidence was her alleged motivation. In addition to seeking the excitement of the code blue alarm and the resuscitation process, plus the recognition for having struggled valiantly to save the patient, it was suggested that she sought to impress her boyfriend, who also worked at the hospital. Moreover, she had access to the epinephrine. But since no one had seen her administer any fatal injections, the case against her, while suggestive, was purely circumstantial. Although the patients involved were mostly middle-aged men not regarded as potential heart attack victims, it was possible that their attacks had occurred naturally. What tipped the balance, and led to a decision to indict Gilbert for multiple murder, was Gehlbach's statistical analysis.

THE SCIENCE OF STATE

Statistics is widely used in law enforcement in many ways and for many purposes. In *NUMB3RS*, Charlie often carries out a statistical analysis, and the use of statistical techniques will appear in many chapters in this book, often without our making explicit mention of the fact. But what exactly does statistics entail? And why was the word in the singular in that last sentence?

The word "statistics" comes from the Latin term *statisticum collegium*, meaning "council of state" and the Italian word *statista*, meaning "statesman," which reflects the initial uses of the technique. The German word *Statistik* likewise originally meant the analysis of data about the state. Until the nineteenth century, the equivalent English term was "political arithmetic," after which the word "statistics" was introduced to refer to any collection and classification of data.

Today, "statistics" really has two connected meanings. The first is the collection and tabulation of data; the second is the use of mathematical and other methods to draw meaningful and useful conclusions from tabulated data. Some statisticians refer to the former activity as "little-s statistics" and the latter activity as "big-S Statistics". Spelled with a lower-case s, the word is treated as plural when it refers to a collection of numbers. But it is singular when used to refer to the activity of collecting and tabulating those numbers. "Statistics" (with a capital S) refers to an activity, and hence is singular.

Though many sports fans and other kinds of people enjoy collecting and tabulating numerical data, the real value of little-s statistics is to provide the data for big-S Statistics. Many of the mathematical techniques used in big-S Statistics involve the branch of mathematics known as probability theory, which began in the sixteenth and seventeenth centuries as an attempt to understand the likely outcomes of games of chance, in order to increase the likelihood of winning. But whereas probability theory is a definite branch of mathematics, Statistics is essentially an applied science that uses mathematical methods.

While the law enforcement profession collects a large quantity of little-s statistics, it is the use of big-S Statistics as a tool in fighting crime that we shall focus on. (From now on we shall drop the "big S", "little s" terminology and use the word "statistics" the way statisticians do, to mean both, leaving the reader to determine the intended meaning from the context.)

Although some applications of statistics in law enforcement use sophisticated methods, the basic techniques covered in a first-semester college statistics course are often enough to crack a case.

This was certainly true for *United States v. Kristen Gilbert*. In that case, a crucial question for the grand jury was whether there were significantly more deaths in the unit when Kristen Gilbert was on duty than at other times. The key word here is "significantly". One or two extra deaths on her watch could be coincidence. How many deaths would it take to reach the level of "significance" sufficient to indict Gilbert? This is a question that only statistics can answer. Accordingly, Stephen Gehlbach was asked to provide the grand jury with a summary of his findings.

HYPOTHESIS TESTING

Gehlbach's testimony was based on a fundamental statistical technique known as hypothesis testing. This method uses probability theory to determine whether an observed outcome is so unusual that it is highly unlikely to have occurred naturally.

One of the first things Gehlbach did was plot the annual number of deaths at the hospital from 1988 through 1997, broken down by shifts—midnight to 8:00 AM, 8:00 AM to 4:00 PM, and 4:00 PM to midnight. The resulting graph is shown in Figure 1. Each vertical bar shows the total number of deaths in the year during that particular shift.

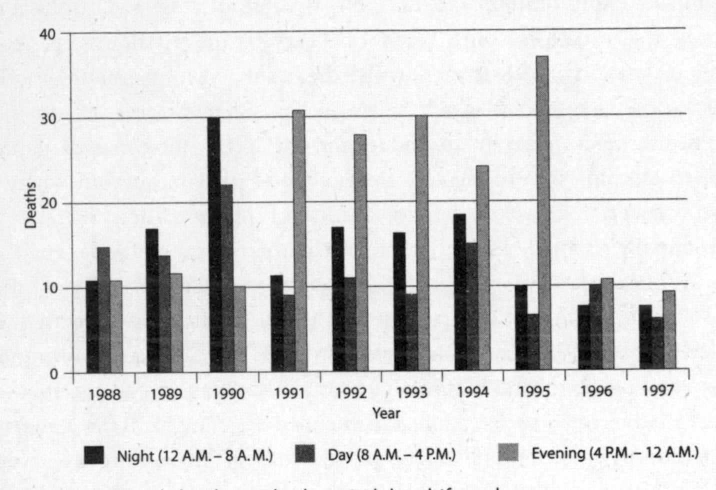

Figure 1. Total deaths at the hospital, by shift and year.

The graph shows a definite pattern. For the first two years, there were around ten deaths per year on each shift. Then, for each of the years 1990 through 1995, one of the three shifts shows between 25 and 35 deaths per year. Finally, for the last two years, the figures drop back to roughly ten deaths on each of the three shifts. When the investigators examined Kristen Gilbert's work record, they discovered that she started work in Ward C in March 1990 and stopped working at the hospital in February 1996. Moreover, for each of the years she worked at the VA, the shift that showed the dramatically increased number of deaths was the one she worked. To a layperson, this might suggest that Gilbert was clearly responsible for the deaths, but on its own it would not be sufficient to secure a conviction—indeed, it might not be enough to justify even an indictment. The problem is that it may be just a coincidence. The job of the statistician in this situation is to determine just how unlikely such a coincidence would be. If the answer is that the likelihood of such a coincidence is, say, 1 in 100, then Gilbert might well be innocent; and even 1 in 1,000 leaves some doubt as to her guilt; but with a likelihood of, say, 1 in 100,000, most people would find the evidence against her to be pretty compelling.

To see how hypothesis testing works, let's start with the simple example of tossing a coin. If the coin is perfectly balanced (i.e., unbiased or fair), then the probability of getting heads is 0.5.* Suppose we toss the coin ten times in a row to see if it is biased in favor of heads. Then we can get a range of different outcomes, and it is possible to compute the likelihood of different results. For example, the probability of getting at least six heads is about 0.38. (The calculation is straightforward but a bit intricate, because there are many possible ways you can get six or more heads in ten tosses, and you have to take account of all of them.) The figure of 0.38 puts a precise numerical value on the fact that, on an intuitive level, we would not be surprised if ten coin tosses gave six or more heads. For at least seven heads, the probability works out at 0.17, a figure that corresponds to our intuition that seven or more heads is somewhat unusual but certainly not a cause for suspicion that the coin was biased. What would surprise us is nine or ten heads, and for that the probability works out at about 0.01, or 1 in 100. The probability of getting ten heads is about 0.001, or 1 in 1,000, and if that happened we would definitely suspect an unfair coin. Thus, by tossing the coin ten times, we can form a reliable, precise judgment, based on mathematics, of the hypothesis that the coin is unbiased.

In the case of the suspicious deaths at the Veteran's Affairs Medical Center, the investigators wanted to know if the number of deaths that occurred when Kristen Gilbert was on duty was so unlikely that it could not be merely happenstance. The math is a bit more complicated than for the coin tossing, but the idea is the same. Table 1 gives the data the investigators had at their disposal. It gives numbers of shifts, classified in different ways, and covers the eighteen-month period ending in February

---

*Actually, this is not entirely accurate. Because of inertial properties of a physical coin, there is a slight tendency for it to resist turning, with the result that, if a perfectly balanced coin is given a random initial flip, the probability that it will land the same way up as it started is about 0.51. But we will ignore this caveat in what follows.

1996, the month when the three nurses told their supervisor of their concerns, shortly after which Gilbert took a medical leave.

GILBERT PRESENT	DEATH ON SHIFT		
	YES	NO	TOTAL
YES	40	217	257
NO	34	1,350	1,384
TOTAL	74	1,567	1,641

Table 1. The data for the statistical analysis in the Gilbert case.

Altogether, there were 74 deaths, spread over a total of 1,641 shifts. If the deaths are assumed to have occurred randomly, these figures suggest that the probability of a death on any one shift is about 74 out of 1,641, or 0.045. Focusing now on the shifts when Gilbert was on duty, there were 257 of them. If Gilbert was not killing any of the patients, we would expect there to be around $0.045 \times 257 = 11.6$ deaths on her shifts, i.e., around 11 or 12 deaths. In fact there were more—40 to be precise. How likely is this? Using mathematical methods similar to those for the coin tosses, statistician Gehlbach calculated that the probability of having 40 or more of the 74 deaths occur on Gilbert's shifts was less than 1 in 100 million. In other words, it is unlikely in the extreme that Gilbert's shifts were merely "unlucky" for the patients.

The grand jury decided there was sufficient evidence to indict Gilbert—presumably the statistical analysis was the most compelling evidence, but we cannot know for sure, as a grand jury's deliberations are not public knowledge. She was accused of four specific murders and three attempted murders. Because the VA is a federal facility, the trial would be in a federal court rather than a state court, and subject to federal laws. A significant consequence of this fact for Gilbert was that although Massachusetts does not have a death penalty, federal law does, and that is what the prosecutor asked for.

STATISTICS IN THE COURTROOM?

An interesting feature of this case is that the federal trial judge ruled in pretrial deliberations that the statistical evidence should not be presented in court. In making his ruling, the judge took note of a submission by a second statistician brought into the case, George Cobb of Mount Holyoke College.

Cobb and Gehlbach did not disagree on any of the statistical analysis. (In fact, they ended up writing a joint article about the case.) Rather, their roles were different, and they were addressing different issues. Gehlbach's task was to use statistics to determine if there were reasonable grounds to suspect Gilbert of multiple murder. More specifically, he carried out an analysis that showed that the increased numbers of deaths at the hospital during the shifts when Gilbert was on duty could not have arisen due to chance variation. That was sufficient to cast suspicion on Gilbert as the cause of the increase, but not at all enough to prove that she *did* cause the increase. What Cobb argued was that the establishment of a statistical relationship does not explain the cause of that relationship. The judge in the case accepted this argument, since the purpose of the trial was not to decide if there were grounds to make Gilbert a suspect—the grand jury and the state attorney's office had done that. Rather, the job before the court was to determine whether or not Gilbert caused the deaths in question. His reason for excluding the statistical evidence was that, as experiences in previous court cases had demonstrated, jurors not well versed in statistical reasoning—and that would be almost all jurors—typically have great difficulty appreciating why odds of 1 in 100 million against the suspicious deaths occurring by chance does *not* imply that the odds that Gilbert did not kill the patients are likewise 1 in 100 million. The original odds could be *caused* by something else.

Cobb illustrated the distinction by means of a famous example from the long struggle physicians and scientists had in overcoming the powerful tobacco lobby to convince governments and the public that cigarette smoking causes lung cancer. Table 2 shows the mortality rates for three categories of people: nonsmokers, cigarette smokers, and cigar and pipe smokers.

Nonsmokers	20.2
Cigarette smokers	20.5
Cigar and pipe smokers	35.3

Table 2. Mortality rates per 1,000 people per year.

At first glance, the figures in Table 2 seem to indicate that cigarette smoking is not dangerous but pipe and cigar smoking are. However, this is not the case. There is a crucial variable lurking behind the data that the numbers themselves do not indicate: age. The average age of the nonsmokers was 54.9, the average age of the cigarette smokers was 50.5, and the average age of the cigar and pipe smokers was 65.9. Using statistical techniques to make allowance for the age differences, statisticians were able to adjust the figures to produce Table 3.

Nonsmokers	20.3
Cigarette smokers	28.3
Cigar and pipe smokers	21.2

Table 3. Mortality rates per 1,000 people per year, adjusted for age.

Now a very different pattern emerges, indicating that cigarette smoking is highly dangerous.

Whenever a calculation of probabilities is made based on observational data, the most that can generally be concluded is that there is a correlation between two or more factors. That can mean enough to spur further investigation, but on its own it does not establish causation. There is always the possibility of a hidden variable that lies behind the correlation.

When a study is made of, say, the effectiveness or safety of a new drug or medical procedure, statisticians handle the problem of hidden parameters by relying not on observational data, but instead by conducting a randomized, double-blind trial. In such a study, the target population is divided into two groups by an entirely random procedure, with the group allocation unknown to both the experimental subjects and the caregivers administering the drug or treatment (hence the term "double-blind"). One group is given the new drug or treatment, the other is given a placebo or dummy treatment. With such an experiment, the random allocation into groups overrides the possible effect of hidden parameters, so that in this case a low probability that a positive result is simply chance variation can indeed be taken as conclusive evidence that the drug or treatment is what caused the result.

In trying to solve a crime, there is of course no choice but to work with the data available. Hence, use of the hypothesis-testing procedure, as in the Gilbert case, can be highly effective in the identification of a suspect, but other means are generally required to secure a conviction.

In *United States v. Kristen Gilbert*, the jury was not presented with Gehlbach's statistical analysis, but they did find sufficient evidence to convict her on three counts of first-degree murder, one count of second-degree murder, and two counts of attempted murder. Although the prosecution asked for the death sentence, the jury split 8–4 on that issue, and accordingly Gilbert was sentenced to life imprisonment with no possibility of parole.

POLICING THE POLICE

Another use of basic statistical techniques in law enforcement concerns the important matter of ensuring that the police themselves obey the law.

Law enforcement officers are given a considerable amount of power over their fellow citizens, and one of the duties of society is to make certain that they do not abuse that power. In particular, police officers are supposed to treat everyone equally and fairly, free of any bias based on gender, race, ethnicity, economic status, age, dress, or religion.

But determining bias is a tricky business and, as we saw in our previous discussion of cigarette smoking, a superficial glance at the statistics can sometimes lead to a completely false conclusion. This is illustrated in a particularly dramatic fashion by the following example, which, while not related to police activity, clearly indicates the need to approach statistics with some mathematical sophistication.

In the 1970s, somebody noticed that 44 percent of male applicants to the graduate school of the University of California at Berkeley were accepted, but only 35 percent of female applicants were accepted. On the face of it, this looked like a clear case of gender discrimination, and, not surprisingly (particularly at Berkeley, long acknowledged as home to many leading advocates for gender equality), there was a lawsuit over gender bias in admissions policies.

It turns out that Berkeley applicants do not apply to the graduate school, but to individual programs of study—such as engineering, physics, or English—so if there is any admissions bias, it will occur within one or more particular program. Table 4 gives the admission data program by program:

Major	Male apps	% admit	Female apps	% admit
A	825	62	108	82
B	560	63	25	68
C	325	37	593	34
D	417	33	375	35
E	191	28	393	24
F	373	6	341	7

Table 4. Admission figures from the University of California at Berkeley on a program-by-program basis.

If you look at each program individually, however, there doesn't appear to be an advantage in admission for male applicants. Indeed, the percentage of female applicants admitted to heavily subscribed program A is considerably higher than for males, and in all other programs the percentages are fairly close. So how can there appear to be an advantage for male applicants overall?

To answer this question, you need to look at what programs males and females applied to. Males applied heavily to programs A and B, females applied primarily to programs C, D, E, and F. The programs that females applied to were more difficult to get into than those for males (the percentages admitted are low for both genders), and this is why it appears that males had an admission advantage when looking at the aggregate data.

There was indeed a gender factor at work here, but it had nothing to do with the university's admissions procedures. Rather, it was one of self-selection by the applying students, where female applicants avoided progams A and B.

The Berkeley case was an example of a phenomenon known as Simpson's paradox, named for E. H. Simpson, who studied this curious phenomenon in a famous 1951 paper.*

HOW DO YOU DETERMINE BIAS?

With the above cautionary example in mind, what should we make of the study carried out in Oakland, California, in 2003 (by the RAND Corporation, at the request of the Oakland Police Department's Racial Profiling Task Force), to determine if there was systematic racial bias in the way police stopped motorists?

The RAND researchers analyzed 7,607 vehicle stops recorded by Oakland police officers between June and December 2003, using various statistical tools to examine a number of variables to uncover any evidence that suggested racial profiling. One figure they found was that blacks were involved in 56 percent of all traffic stops studied, although they make up just 35 percent of Oakland's residential population. Does this finding indicate racial profiling? Well, it might, but as soon as you look more closely at what other factors could be reflected in those numbers, the issue is by no means clear cut.

For instance, like many inner cities, Oakland has some areas with much higher crime rates than others, and the police patrol those higher crime areas at a much greater rate than they do areas having less crime. As a result, they make more traffic stops in those areas. Since the higher crime areas typically have greater concentrations of minority groups, the higher rate of traffic stops in those areas manifests itself as a higher rate of traffic stops of minority drivers.

To overcome these uncertainties, the RAND researchers devised a particularly ingenious way to look for possible racial bias. If racial profiling was occurring, they reasoned, stops of minority drivers would be higher when the officers could determine the driver's race prior to making the stop. Therefore, they compared the stops made during a period

*E. H. Simpson. "The Interpretation of Interaction in Contingency Tables," *Journal of the Royal Statistical Society*, Ser. B, 13 (1951) 238–241.

just before nightfall with those made after dark—when the officers would be less likely to be able to determine the driver's race. The figures showed that 50 percent of drivers stopped during the daylight period were black, compared with 54 percent when it was dark. Based on that finding, there does not appear to be systematic racial bias in traffic stops.

But the researchers dug a little further, and looked at the officers' own reports as to whether they could determine the driver's race prior to making the stop. When officers reported knowing the race in advance of the stop, 66 percent of drivers stopped were black, compared with only 44 percent when the police reported not knowing the driver's race in advance. This is a fairly strong indicator of racial bias.*

---

*Sadly, despite many efforts to eliminate the problem, racial bias by police seems to be a persistent issue throughout the country. To cite just one recent report, *An Analysis of Traffic Stop Data in Riverside, California*, by Larry K. Gaines of the California State University in San Bernardino, published in *Police Quarterly*, 9, 2, June 2006, pp. 210–233: "The findings from racial profiling or traffic stop studies have been fairly consistent: Minorities, especially African Americans, are stopped, ticketed, and searched at a higher rate as compared to Whites. For example, Lamberth (cited in *State v. Pedro Soto*, 1996) found that the Maryland State Police stopped and searched African Americans at a higher rate as compared to their rate of speeding violations. Harris (1999) examined court records in Akron, Dayton, Toledo, and Columbus, Ohio, and found that African Americans were cited at a rate that surpassed their representation in the driving population. Cordner, Williams, and Zuniga (2000) and Cordner, Williams, and Velasco (2002) found similar trends in San Diego, California. Zingraff and his colleagues (2000) examined stops by the North Carolina Highway Patrol and found that African Americans were overrepresented in stops and searches."

CHAPTER

3 Data Mining

Finding Meaningful Patterns in Masses of Information

BRUTUS

Charlie Eppes is sitting in front of a bank of computers and television monitors. He is testing a computer program he is developing to help police monitor large crowds, looking for unusual behavior that could indicate a pending criminal or terrorist act. His idea is to use standard mathematical equations that describe the flow of fluids—in rivers, lakes, oceans, tanks, pipes, even blood vessels.* He is trying out the new system at a fund-raising reception for one of the California state senators. Overhead cameras monitor the diners as they move around the room, and Charlie's computer program analyzes the "flow" of the people. Suddenly the test takes on an unexpected aspect. The FBI receives a telephone warning that a gunman is in the room, intending to kill the senator.

The software works, and Charlie is able to identify the gunman, but Don and his team are not able to get to the killer before he has shot the senator and then turned the gun on himself.

The dead assassin turns out to be a Vietnamese immigrant, a former Vietcong member, who, despite having been in prison in California,

*The idea is based on several real-life projects to use the equations that describe fluid flows in order to analyze various kinds of crowd activity, including freeway traffic flow, spectators entering and leaving a large sports stadium, and emergency exits from burning buildings.

somehow managed to obtain U.S. citizenship and be the recipient of a regular pension from the U.S. Army. He had also taken the illegal drug speed on the evening of the assassination. When Don makes some enquiries to find out just what is going on, he is visited by a CIA agent who asks for help in trying to prevent too much information about the case leaking out. Apparently the dead killer had been part of a covert CIA behavior modification project carried out in California prisons during the 1960s to turn inmates into trained assassins who, when activated, would carry out their assigned task before killing themselves. (Sadly, this idea is no less fanciful than that of Charlie using fluid flow equations to study crowd behavior.)

But why had this particular individual suddenly become active and murdered the state senator?

The picture becomes much clearer when a second murder occurs. The victim this time is a prominent psychiatrist, the killer a Cuban immigrant. The killer had also spent time in a California prison, and he too was the recipient of regular Army pension checks. But on this occasion, when the assassin tries to shoot himself after killing the victim, the gun fails to go off and he has to flee the scene. A fingerprint identification from the gun soon leads to his arrest.

When Don realizes that the dead senator had been urging a repeal of the statewide ban on the use of behavior modification techniques on prison inmates, and that the dead psychiatrist had been recommending the re-adoption of such techniques to overcome criminal tendencies, he quickly concludes that someone has started to turn the conditioned assassins on the very people who were pressing for the reuse of the techniques that had produced them. But who?

Don thinks his best line of investigation is to find out who supplied the guns that the two killers had used. He knows that the weapons originated with a dealer in Nevada. Charlie is able to provide the next step, which leads to the identification of the individual behind the two assassinations. He obtains data on all gun sales involving that particular dealer and analyzes the relationships among all sales that originated there. He explains that he is employing mathematical techniques similar to those used to analyze calling patterns on the telephone network—an approach used frequently in real-life law enforcement.

This is what viewers saw in the third-season episode of *NUMB3RS* called "Brutus" (the code name for the fictitious CIA conditioned-assassinator project), first aired on November 24, 2006. As usual, the mathematics Charlie uses in the show is based on real life.

The method Charlie uses to track the gun distribution is generally referred to as "link analysis," and is one among many that go under the collective heading of "data mining." Data mining obtains useful information among the mass of data that is available—often publicly—in modern society.

FINDING MEANING IN INFORMATION

Data mining was initially developed by the retail industry to detect customer purchasing patterns. (Ever wonder why supermarkets offer customers those loyalty cards—sometimes called "club" cards—in exchange for discounts? In part it's to encourage customers to keep shopping at the same store, but low prices would do that. The significant factor for the company is that it enables them to track detailed purchase patterns that they can link to customers' home zip codes, information that they can then analyze using data-mining techniques.)

Though much of the work in data mining is done by computers, for the most part those computers do not run autonomously. Human expertise also plays a significant role, and a typical data-mining investigation will involve a constant back-and-forth interplay between human expert and machine.

Many of the computer applications used in data mining fall under the general area known as artificial intelligence, although that term can be misleading, being suggestive of computers that think and act like people. Although many people believed that was a possibility back in the 1950s when AI first began to be developed, it eventually became clear that this was not going to happen within the foreseeable future, and may well never be the case. But that realization did not prevent the development of many "automated reasoning" programs, some of which eventually found a powerful and important use in data mining, where the human expert often provides the "high-level intelligence" that guides the computer programs that do the bulk of the work. In this way, data mining provides an excellent example of the power that results when human brains team up with computers.

Among the more prominent methods and tools used in data mining are:

- *Link analysis*—looking for associations and other forms of connection among, say, criminals or terrorists
- *Geometric clustering*—a specific form of link analysis
- *Software agents*—small, self-contained pieces of computer code that can monitor, retrieve, analyze, and act on information
- *Machine learning*—algorithms that can extract profiles of criminals and graphical maps of crimes
- *Neural networks*—special kinds of computer programs that can predict the probability of crimes and terrorist attacks.

We'll take a brief look at each of these topics in turn.

LINK ANALYSIS

Newspapers often refer to link analysis as "connecting the dots." It's the process of tracking connections between people, events, locations, and organizations. Those connections could be family ties, business relationships, criminal associations, financial transactions, in-person meetings, e-mail exchanges, and a host of others. Link analysis can be particularly powerful in fighting terrorism, organized crime, money laundering ("follow the money"), and telephone fraud.

Link analysis is primarily a human-expert driven process. Mathematics and technology are used to provide a human expert with powerful, flexible computer tools to uncover, examine, and track possible connections. Those tools generally allow the analyst to represent linked data as a network, displayed and examined (in whole or in part) on the computer screen, with nodes representing the individuals or organizations or locations of interest and the links between those nodes representing relationships or transactions. The tools may also allow the analyst to investigate and record details about each link, and to discover new nodes that connect to existing ones or new links between existing nodes.

For example, in an investigation into a suspected crime ring, an investigator might carry out a link analysis of telephone calls a suspect has made or received, using telephone company call-log data, looking at factors such as number called, time and duration of each call, or number called next. The investigator might then decide to proceed further along the call network, looking at calls made to or from one or more of the individuals who had had phone conversations with the initial suspect. This process can bring to the investigator's attention individuals not previously known. Some may turn out to be totally innocent, but others could prove to be criminal collaborators.

Another line of investigation may be to track cash transactions to and from domestic and international bank accounts.

Still another line may be to examine the network of places and people visited by the suspect, using such data as train and airline ticket purchases, points of entry or departure in a given country, car rental records, credit card records of purchases, websites visited, and the like.

Given the difficulty nowadays of doing almost anything without leaving an electronic trace, the challenge in link analysis is usually not one of having insufficient data, but rather of deciding which of the megabytes of available data to select for further analysis. Link analysis works best when backed up by other kinds of information, such as tips from police informants or from neighbors of possible suspects.

Once an initial link analysis has identified a possible criminal or terrorist network, it may be possible to determine who the key players are by examining which individuals have the most links to others in the network.

GEOMETRIC CLUSTERING

Because of resource limitations, law enforcement agencies generally focus most of their attention on major crime, with the result that minor offenses such as shoplifting or house burglaries get little attention. If, however, a single person or an organized gang commits many such crimes on a regular basis, the aggregate can constitute significant criminal activity that deserves greater police attention. The problem facing the authorities, then, is to identify within the large numbers of minor crimes that take place every day, clusters that are the work of a single individual or gang.

One example of a "minor" crime that is often carried out on a regular basis by two (and occasionally three) individuals acting together is the so-called *bogus official burglary* (or *distraction burglary*). This is where two people turn up at the front door of a homeowner (elderly people are often the preferred targets) posing as some form of officials—perhaps telephone engineers, representatives of a utility company, or local government agents—and, while one person secures the attention of the homeowner, the other moves quickly through the house or apartment taking any cash or valuables that are easily accessible.

Victims of bogus official burglaries often file a report to the police, who will send an officer to the victim's home to take a statement. Since the victim will have spent considerable time with one of the perpetrators (the distracter), the statement will often include a fairly detailed description—gender, race, height, body type, approximate age, general facial appearance, eyes, hair color, hair length, hair style, accent, identifying physical marks, mannerisms, shoes, clothing, unusual jewelry, etc.—together with the number of accomplices and their genders. In principle, this wealth of information makes crimes of this nature ideal for data mining, and in particular for the technique known as *geometric clustering*, to identify groups of crimes carried out by a single gang. Application of the method is, however, fraught with difficulties, and to date the method appears to have been restricted to one or two experimental studies. We'll look at one such study, both to show how the method works and to illustrate some of the problems often faced by the data-mining practitioner.

The following study was carried out in England in 2000 and 2001 by researchers at the University of Wolverhampton, together with the West Midlands Police.* The study looked at victim statements from bogus official burglaries in the police region over a three-year period. During that period, there were 800 such burglaries recorded, involving

*Ref. R. Adderley and P. B. Musgrove, General Review of Police Crime Recording and Investigation Systems, *Policing: An International Journal of Police Strategies and Management*, 24 (1), 2001, pp.110–114.

1,292 offenders. This proved to be too great a number for the resources available for the study, so the analysis was restricted to those cases where the distracter was female, a group comprising 89 crimes and 105 offender descriptions.

The first problem encountered was that the descriptions of the perpetrators was for the most part in narrative form, as written by the investigating officer who took the statement from the victim. A data-mining technique known as text mining had to be used to put the descriptions into a structured form. Because of the limitations of the text-mining software available, human input was required to handle many of the entries; for instance, to cope with spelling mistakes, ad hoc or inconsistent abbreviations (e.g., "Bham" or "B'ham" for "Birmingham"), and the use of different ways of expressing the same thing (e.g., "Birmingham accent", "Bham accent", "local accent", "accent: local", etc.).

After some initial analysis, the researchers decided to focus on eight variables: age, height, hair color, hair length, build, accent, race, and number of accomplices.

Once the data had been processed into the appropriate structured format, the next step was to use geometric clustering to group the 105 offender descriptions into collections that were likely to refer to the same individual. To understand how this was done, let's first consider a method that at first sight might appear to be feasible, but which soon proves to have significant weaknesses. Then, by seeing how those weaknesses may be overcome, we will arrive at the method used in the British study.

First, you code each of the eight variables numerically. Age—often a guess—is likely to be recorded either as a single figure or a range; if it is a range, take the mean. Gender (not considered in the British Midlands study because all the cases examined had a female distracter) can be coded as 1 for male, 0 for female. Height may be given as a number (inches), a range, or a term such as "tall", "medium", or "short"; again, some method has to be chosen to convert each of these to a single figure. Likewise, schemes have to be devised to represent each of the other variables as a number.

When the numerical coding has been completed, each perpetrator description is then represented by an eight-vector, the coordinates of a point in eight-dimensional geometric (Euclidean) space. The familiar distance measure of Euclidean geometry (the Pythagorean metric) can then be used to measure the geometric distance between each pair of points. This gives the distance between two vectors $(x_1, \ldots, x_8)$ and $(y_1, \ldots, y_8)$ as:

$$\sqrt{[(x_1 - y_1)^2 + \ldots + (x_8 - y_8)^2]}$$

Points that are close together under this metric are likely to correspond to perpetrator descriptions that have several features in common; and the closer the points, the more features the descriptions are likely to have in common. (Remember, there are problems with this approach, which we'll get to momentarily. For the time being, however, let's suppose that things work more or less as just described.)

The challenge now is to identify clusters of points that are close together. If there were only two variables, this would be easy. All the points could be plotted on a single x,y-graph and visual inspection would indicate possible clusters. But human beings are totally unable to visualize eight-dimensional space, no matter what assistance the software system designers provide by way of data visualization tools. The way around this difficulty is to reduce the eight-dimensional array of points (descriptions) to a two-dimensional array (i.e., a matrix or table). The idea is to arrange the data points (that is, the vector representatives of the offender descriptions) in a two-dimensional grid in such a way that:

1. pairs of points that are extremely close together in the eight-dimensional space are put into the same grid entry;

2. pairs of points that are neighbors in the grid are close together in the eight-dimensional space; and

3. points that are farther apart in the grid are farther apart in the space.

This can be done using a special kind of computer program known as a neural net, in particular, a Kohonen self-organizing map (or SOM).

Neural nets (including SOMs) are described later in the chapter. For now, all we need to know is that these systems, which work iteratively, are extremely good at homing in (over the course of many iterations) on patterns, such as geometric clusters of the kind we are interested in, and thus can indeed take an eight-dimensional array of the kind described above and place the points appropriately in a two-dimensional grid. (Part of the skill required to use an SOM effectively in a case such as this is deciding in advance, or by some initial trial and error, what are the optimal dimensions of the final grid. The SOM needs that information in order to start work.)

Once the data has been put into the grid, law enforcement officers can examine grid squares that contain several entries, which are highly likely to come from a single gang responsible for a series of crimes, and can visually identify clusters on the grid, where there is also a likelihood that they represent gang activity. In either case, the officers can examine the corresponding original crime statement entries, looking for indications that those crimes are indeed the work of a single gang.

Now let's see what goes wrong with the method just described, and how to correct it.

The first problem is that the original encoding of entries as numbers is not systematic. This can lead to one variable dominating others when the entries are clustered using geometric distance (the Pythagorean metric) in eight-dimensional space. For example, a dimension that measures height (which could be anything between 60 inches and 76 inches) would dominate the entry for gender (0 or 1). So the first step is to scale (in mathematical terminology, *normalize*) the eight numerical variables, so that each one varies between 0 and 1.

One way to do that would be to simply scale down each variable by a multiplicative scaling factor appropriate for that particular feature (height, age, etc.). But that will introduce further problems when the separation distances are calculated; for example, if gender and height are among the variables, then, all other variables being roughly the same, a very tall woman would come out close to a very short man (because female gives a 0 and male gives a 1, whereas tall comes out close to 1 and short close to 0). Thus, a more sophisticated normalization procedure has to be used.

The approach finally adopted in the British Midlands study was to make every numerical entry binary (just 0 or 1). This meant splitting the continuous variables (age and height) into overlapping ranges (a few years and a few inches, respectively), with a 1 denoting an entry in a given range and a 0 meaning outside that range, and using pairs of binary variables to encode each factor of hair color, hair length, build, accent, and race. The exact coding chosen was fairly specific to the data being studied, so there is little to be gained from providing all the details here. (The age and height ranges were taken to be overlapping to account for entries toward the edges of the chosen ranges.) The normalization process resulted in a set of 46 binary variables. Thus, the geometric clustering was done over a geometric space of 46 dimensions.

Another problem was how to handle missing data. For example, what do you do if a victim's statement says nothing about the perpetrator's accent? If you enter a 0, that would amount to assigning an accent. But what will the clustering program do if you leave that entry blank? (In the British Midlands study, the program would treat a missing entry as 0.) Missing data points are in fact one of the major headaches for data miners, and there really is no universally good solution. If there are only a few such cases, you could either ignore them or else see what solutions you get with different values entered.

As mentioned earlier, a key decision that has to be made before the SOM can be run is the size of the resulting two-dimensional grid. It needs to be small enough so that the SOM is forced to put some data points into the same grid squares, and will also result in some non-empty grid squares having non-empty neighbors. The investigators in the British Midlands study eventually decided to opt for a five-by-seven grid. With 105 offender descriptions, this forced the SOM to create several multi-entry clusters.

The study itself concluded with experienced police officers examining the results and comparing them with the original victim statements and other relevant information (such as geographic proximity of crimes over a short timespan, which would be another indicator of a gang activity, not used in the cluster analysis), to determine how well the process performed. Though all parties involved in the study declared it to be successful, the significant amount of person-hours required means that such methods need further development, and greater automation of the various steps, before they can become widely used to fight criminal activity of the kind the study focused on. However, the method can be used to detect clusters in other kinds of criminal activity, such as terrorism. In such cases, where the stakes are so high, it may be well worth the investment of personnel and resources to make the method work.

SOFTWARE AGENTS

Software agents, a product of AI research, are essentially self-contained (and, in general, relatively small) computer programs designed to achieve specific goals, and that act autonomously, responding to changes in the environment in which they operate. Their autonomy is a result of their incorporating a range of different actions they can take, depending on particular inputs. Put crudely, they include a large number of if/then instructions.

For example, FinCEN, the U.S. Treasury agency whose job it is to detect money laundering, reviews every cash transaction involving more than $10,000. As there are about 10 million such transactions each year, this cannot be done manually. Instead, the agency uses software agents to carry out the monitoring automatically, using link analysis, among other tools, to look for unusual activity that might indicate fraud.

Banks use software agents to monitor credit card activity, looking for an unusual spending pattern that might indicate a stolen card. (You may have experienced having your credit card rejected when you tried to use it in novel circumstances, such as overseas or else in a city or a foreign country where there had been—most likely unbeknownst to you—recent fraudulent credit card use.)

The Defense Department, among other government and non-government organizations, has invested large amounts of money in the development of software agents for intelligence gathering and analysis. Typically, the strategy is to develop a coordinated system of agents that communicate with one another, each of which is designed to carry out one particular subtask. For example, a coordinated surveillance system to provide an early warning of a biological attack might include the following:

- Agents that receive and correlate data from different databases
- Agents that extract potentially relevant data from those databases
- Agents that analyze selected data and look for unusual patterns of biological events
- Agents that classify abnormalities and identify specific pathogens
- Agents that provide alerts to the emergency response personnel.

The initial data examined might include physicians' reports or patient symptoms, hospital outpatient reports, school attendance records, or sales of particular drugs by pharmacies. In each case, a sudden change from an established pattern might be due to a naturally occurring epidemic, but could provide the first signs of a biological attack. Humans would be unable to summarize the masses of data and survey the results in order to detect a changing situation sufficiently quickly to be able to initiate countermeasures. This has to be done using software.

MACHINE LEARNING

Machine learning, another branch of artificial intelligence, is perhaps the single most important tool within the law enforcement community's data-mining arsenal when it comes to profiling (and hence, one hopes, catching or preventing) criminals and terrorists.

Much of the power of machine learning algorithms stems from the fact that they automate the process of searching for and identifying key features in masses of data. This is something that a trained person can do—usually better, actually—but only for small quantities of data. Machine learning algorithms are capable of finding the proverbial needle in a haystack.

For example, if you wanted to uncover a set of features that are characteristic of a terrorist or drug smuggler, you could apply an appropriate machine learning system—of which there are many commercially available—to a database of known (that is, already caught) terrorists or drug smugglers.

Following some initial input from you to determine the range of possible characteristics, the software would quiz the database in much the same fashion as in the familiar twenty-questions parlor game. The output from this process could be a list of if/then conditions, each one with an associated probability estimate, that provide the basis for a program—perhaps to be used at a border crossing—that will check suspects to see if they are likely to be smuggling drugs. Alternatively, the database quizzing process might generate a decision tree that likewise may be used as the basis for a program that alerts law enforcement agents to possible terrorists or drug smugglers.

The first stage of this process is most easily understood using a simple example. Suppose you wanted the machine learning system to predict whether a given item is an apple, an orange, or a banana. You might start by telling it to look at weight, shape, or color. The system looks through its list of appropriate items—in this case, fruit—and first checks weights. It discovers that this feature does not distinguish between the three fruit. It then checks its list against shape. This feature is able to distinguish bananas from the other two (cylindrical/curved, as opposed to spherical), but is not sufficient to identify the fruit in every case. When presented with a test item, checking against shape would give the output

BANANA—100%

if the item is a banana, but

APPLE—50%   ORANGE—50%

in the other cases. Finally, the system checks color. This time it finds that the feature distinguishes the three fruits with 100 percent accuracy.

When a machine learning algorithm is run against a sufficiently large database of past examples, it can often generate a short checklist or decision tree that a border guard or law enforcement agent, faced with a possible criminal or terrorist, can instruct the system to run through in real time to determine possible or likely guilt. Based on the aggregate probability of the suspect's guilt, the system can even advise the agent on what action to take, from "let through" to "arrest immediately".

For instance, although the actual systems used are not made public, it seems highly likely that an individual trying to enter the country would be held for further questioning if he or she had the following characteristics:

AGE:	20–25
GENDER:	Male
NATIONALITY:	Saudi Arabia
COUNTRY OF RESIDENCE:	Germany
VISA STATUS:	Student
UNIVERSITY:	Unknown
# TIMES ENTERING THE COUNTRY IN THE PAST YEAR:	3
COUNTRIES VISITED DURING THE PAST THREE YEARS:	U.K., Pakistan
FLYING LESSONS:	Yes

The system would probably simply suggest that the agent investigate further based on the first seven features, but the final two would likely trigger more substantive action. (One can imagine the final feature being activated only when several of the earlier ones raise the likelihood that the individual is a terrorist.)

Of course, the above example is grossly simplified to illustrate the general idea. The power of machine learning is that it can build up fairly complex profiles that would escape a human agent. Moreover, using Bayesian methods (see Chapter 6) for updating probabilities, the system can attach a probability to each conclusion. In the above example, the profile might yield the advice:

ASSESSMENT:	Possible terrorist (probability 29%)
ACTION:	Detain and report

Though our example is fictitious, machine learning systems are in daily use by border guards and law enforcement agencies when screening people entering the country for possible drug-smuggling or terrorist activities. Detecting financial fraud is another area where law enforcement agencies make use of machine learning. And the business world also makes extensive use of such systems, in marketing, customer profiling, quality control, supply chain management, distribution, and so forth, while major political parties use them to determine where and how to target their campaigns.

In some applications, machine learning systems operate like the ones described above; others make use of neural networks, which we consider next.

NEURAL NETWORKS

On June 12, 2006, *The Washington Post* carried a full-page advertisement from Visa Corporation, announcing that their record of credit card fraud was near its all-time low, citing neural networks as the leading security measure that the company had taken to stop credit card fraud. Visa's success came at the end of a long period of development of neural network–based fraud prevention measures that began in 1993, when the company was the first to experiment with the use of such systems to reduce the incidence of card fraud. The idea was that by analyzing typical card usage patterns, a neural network–based risk management tool would notify banks immediately when any suspicious activity occurred, so they could inform their customers if a card appears to have been used by someone other than the legitimate cardholder.

Credit card fraud detection is just one of many applications of data mining that involve the use of a neural network. What exactly are neural networks and how do they work?

A neural network is a particular kind of computer program, originally developed to try to mimic the way the human brain works. It is essentially a computer simulation of a complex circuit through which electric current flows. (See Figure 2.)

Neural networks are particularly suited to recognizing patterns, and were introduced into the marketplace in the 1980s, for tasks such as classifying loan applications as good or bad risks, distinguishing legal from fraudulent financial transactions, identifying possible credit card theft, recognizing signatures, and identifying purchasing patterns in branch supermarkets. Law enforcement agencies started using neural networks

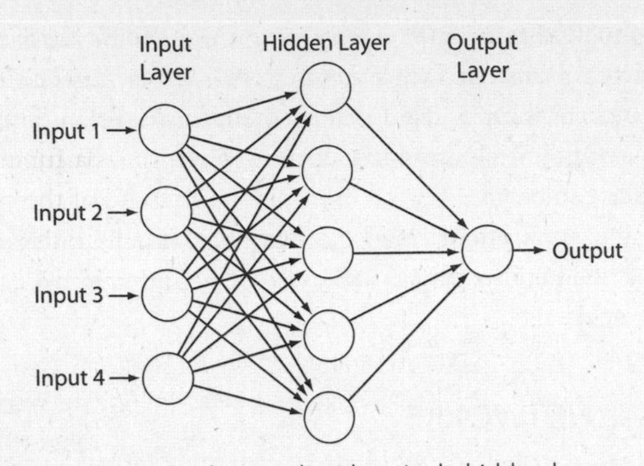

Figure 2. A simple neural network with a single hidden layer and one output node.

soon afterward, applying them to such tasks as recognizing a "forensic fingerprint" that indicates that different cases of arson are likely the work of a single individual, or to recognize activity and behavioral patterns that indicate possible smuggling or terrorist intent.

To go into a little more detail about the technology, a neural network consists of* many (typically several hundred or several thousand) nodes arranged in two or more "parallel layers," with each node in one layer connected to one or more nodes in the adjacent layer. One end-layer is the input layer, the other end-layer is the output layer. All the other layers are called intermediate layers or hidden layers. (The brain-modeling idea is that the nodes simulate neurons and the connections dendrites.) Figure 2 gives the general idea, although a network with so few nodes would be of little practical use.

The network commences an operation cycle when a set of input signals is fed into the nodes of the input layer. Whenever a node anywhere in the network receives an input signal, it sends output signals to all those nodes on the next layer to which it is connected. The cycle completes when signals have propagated through the entire network and an output signal (or signals) emerges from the output node (or the multiple nodes in the output layer if that is how the network is structured). Each

*It's actually more accurate to say "can be regarded as" rather than "consists of," since the entire "neural network" is simulated on a normal digital computer.

input signal and each signal that emerges from a node has a certain "signal strength" (expressed by a number between 1 and 100). Each internode connection has a "transmission strength" (also a number), and the strength of the signal passing along a connection is a function of the signal at the start node and the transmission strength of the connection. Every time a signal is transmitted along a connection, the strength of that connection (also often called its "weight") is increased or decreased proportional to the signal strength, according to a preset formula. (This corresponds to the way that, in a living brain, life experiences result in changes to the strengths of the synaptic connections between neurons in the brain.) Thus, the overall connection-strength configuration of the network changes with each operational cycle.

To use the network to carry out a particular computational task, the input(s) to the computation must be encoded as a set of input signals to the input layer and the corresponding output signal(s) interpreted as a result of the computation. The behavior of the network—what it does to the input(s)—is dependent on the weights of the various network connections. Essentially, the patterns of those weights constitute the network's "memory." The ability of a neural network to perform a particular task at any moment in time depends upon the actual architecture of the network and its current memory.

TRAINING A NEURAL NETWORK

Neural networks are not programmed in the usual sense of programming a computer. In the majority of cases, particularly neural networks used for classification, the application of a network must be preceded by a process of "training" to set the various connection weights.

By way of an example, suppose a bank wanted to train a neural network to recognize unauthorized credit card use. The bank first presents the network with a large number of previous credit card transactions (recorded in terms of user's home address, credit history, spending limit, expenditure, date, amount, location, etc.), each known to be either authentic or fraudulent. For each one, the network has to make a prediction concerning the transaction's authenticity. If the connection weights in the network are initially set randomly or in some neutral way, then some of its predictions will be correct and others wrong. During the training process, the network is "rewarded" each time its prediction is correct and "punished" each time it is wrong. (That is to say, the network is constructed so that a "correct grade"—i.e., positive feedback on its prediction—causes it to continue adjusting the connection weights as before, whereas a "wrong grade" causes it to adjust them differently.) After many cycles (thousands or more), the connection weights will adjust so that on the majority of occasions (generally the *vast* majority) the decision made by the network is correct. What happens is that, over the course of many training cycles, the connection weights in the network will adjust in a way that corresponds to the profiles of legitimate and fraudulent credit card use, whatever those profiles may be (and, of great significance, without the programmer having to know them).

Some skill is required to turn these general ideas into a workable system, and many different network architectures have been developed to build systems that are suited to particular classification tasks.

After completion of a successful training cycle, it can be impossible for a human operator to figure out just what patterns of features (to continue with our current example) of credit card transactions the network has learned to identify as indicative of fraud. All that the operator can know is that the system is accurate to a certain degree of error, giving a correct prediction perhaps 95 percent of the time.

A similar phenomenon can occur with highly trained, highly experienced human experts in a particular domain, such as physicians. An experienced doctor will sometimes examine a patient and say with some certainty what she believes is wrong with the individual, and yet be unable to explain exactly just what specific symptoms led her to make that conclusion.

Much of the value of neural networks comes from the fact that they can acquire the ability to discern feature-patterns that no human could uncover. To take one example, typically just one credit card transaction among every 50,000 is fraudulent. No human could monitor that amount of activity to identify the frauds.

On occasion, however, the very opacity of neural networks—the fact that they can uncover patterns that the human would not normally recognize as such—can lead to unanticipated results. According to one oft-repeated story, some years ago the U.S. Army trained a neural network to recognize tanks despite their being painted in camouflage colors to blend in with the background. The system was trained by showing it many photographs of scenes, some with tanks in, others with no tanks. After many training cycles, the network began to display extremely accurate tank recognition capacity. Finally, the day came to test the system in the field, with real tanks in real locations. And to everyone's surprise, it performed terribly, seeming quite unable to distinguish between a scene with tanks and one without. The red-faced system developers retreated to their research laboratory and struggled to find out what had gone wrong. Eventually, someone realized what the problem was. The photos used to train the system had been taken on two separate days. The photos with tanks in them had been taken on a sunny day, the tank-free photos on a cloudy day. The neural network had certainly learned the difference between the two sets of photos, but the pattern it had discerned had nothing to do with the presence or absence of tanks; rather, the system had learned to distinguish a sunny day scene from a cloudy day scene. The moral of this tale being, of course, that you have to be careful when interpreting exactly which pattern a neural network has identified. That caution aside, however, neural networks have proved themselves extremely useful both in industry and commerce, and in law enforcement and defense.

Various network architectures have been developed to speed up the initial training process before a neural network can be put to work, but in most cases it still takes some time to complete. The principal exceptions are the Kohonen networks (named after Dr. Tevo Kohonen, who developed the idea), also known as Self-Organizing Maps (SOMs), which are used to identify clusters, and which we mentioned in Chapter 3 in connection with clustering crimes into groups that are likely to be the work of one individual or gang.

Kohonen networks have an architecture that incorporates a form of distance measurement, so that they essentially train themselves, without the need for any external feedback. Because they do not require feedback, there is no need for a large body of prior data; they train themselves by cycling repeatedly through the application data. Nevertheless, they function by adjusting connection weights, just like the other, more frequently used neural networks.

One advantage of neural networks over other data-mining systems is that they are much better able to handle the inevitable problem of missing data points that comes with any large body of human-gathered records.

CRIME DATA MINING USING NEURAL NETWORKS

Several commercial systems have been developed to help police solve—and on occasion even stop—crimes.

One such is the Classification System for Serial Criminal Patterns (CSSCP), developed by computer scientists Tom Muscarello and Kamal Dahbur at DePaul University in Chicago. CSSCP sifts through all the case records available to it, assigning numerical values to different aspects of each crime, such as the kind of offence, the perpetrator's sex, height, and age, and the type of weapon or getaway vehicle used. From these figures it builds a crime description profile. A Kohonen-type neural network program then uses this to seek out crimes with similar profiles. If it finds a possible link between two crimes, CSSCP compares when and where they took place to find out whether the same criminals would have had enough time to travel from one crime scene to the other. In a laboratory trial of the system, using three years' worth of data on armed robbery, the system was able to spot ten times as many patterns as a team of experienced detectives with access to the same data.

Another such program is CATCH, which stands for Computer Aided Tracking and Characterization of Homicides. CATCH was developed by Pacific Northwest National Laboratory for the National Institute of Justice and the Washington State Attorney General's Office. It is meant to help law enforcement officials determine connections and relationships in data from ongoing investigations and solved cases. CATCH was built around Washington state's Homicide Investigation Tracking system, which contains the details of 7,000 murders and 6,000 sexual assault cases in the Northwest. CATCH uses a Kohonen-style neural network to cluster crimes through the use of parameters such as modus operandi and signature characteristics of the offenders, allowing analysts to compare one case with similar cases in the database. The system learns about an existing crime, the location of the crime, and the particular characteristics of the offense. The program is subdivided into different tools, each of which places an emphasis on a certain characteristic or group of characteristics. This allows the user to remove certain characteristics which humans determine are unrelated.

Then there is the current particular focus on terrorism. According to the cover story in *BusinessWeek* on August 8, 2005: "Since September 11 more than 3,000 Al Qaeda operatives have been nabbed, and some 100 terrorist attacks have been blocked worldwide, according to the FBI. Details on how all this was pulled off are hush-hush. But no doubt two keys were electronic snooping—using the secret Echelon network—and computer data mining."

Echelon is the global eavesdropping system run by the National Security Agency (NSA) and its counterparts in Canada, Britain, Australia, and New Zealand. The NSA's supercomputers sift through the flood of data gathered by Echelon to spot clues to terrorism planning. Documents the system judges to merit attention go to human translators and analysts, and the rest is dumped. Given the amount of data involved, it's hardly surprising that the system sometimes outperforms the human analysts, generating important information too quickly for humans to examine. For example, two Arabic messages collected on September 10, 2001, hinting of a major event to occur on the next day, were not translated until September 12. (Since that blackest of black days, knowledgeable sources claim that the translation delay has diminished to about twelve hours. The goal, of course, is near-real-time analysis.)

The ultimate goal is the development of data-mining systems that can look through multiple databases and spot correlations that warn of plots being hatched. The Terrorism Information Awareness (TIA) project was supposed to do that, but Congress killed it in 2003 because of privacy concerns. In addition to inspecting multiple commercial and government databases, TIA was designed to spin out its own terrorist scenarios—such as an attack on New York Harbor—and then determine effective means to uncover and blunt the plots. For instance, it might have searched customer lists of diving schools and firms that rent scuba gear, and then looked for similar names on visa applications or airline passenger lists.

I KNOW THAT FACE

Facial recognition systems often make use of neural networks. Current recognition systems reduce the human face to a sequence of numbers (sometimes called a "face print" or a "feature vector"). These numbers are distance measurements at and between pairs of eighty so-called nodal points, key features of the face such as the centers of the eyes, the depths of the eye sockets, cheekbones, jaw line, chin, the width of the nose, and the tip of the nose. (See Figure 3.) Using fast computers, it is possible to compute the face print of a target individual and compare it to the face prints in a database within a few seconds. The comparison cannot be exact, since the angle of observation of the target will be different from that of each photograph used to generate the face print in the database, although this effect can be overcome in part by means of some elementary trigonometric calculations. But this is the kind of "closest match" comparison task that neural networks can handle well.

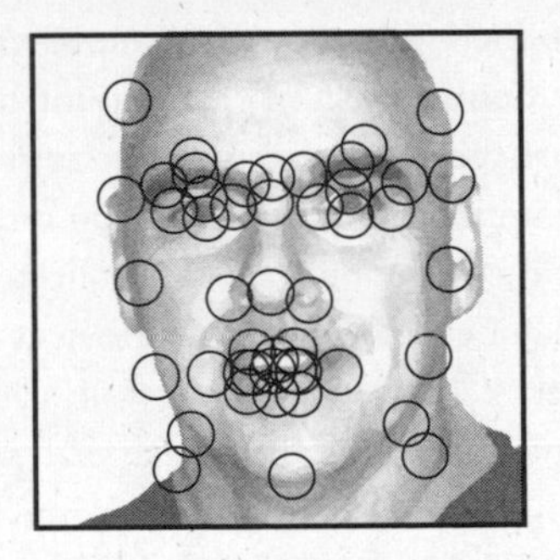

Figure 3. Many facial recognition systems are based on measurements of and between key locations on the face called nodal points.

One advantage of facial recognition using neural network comparisons of face prints is that it is not affected by surface changes such as wearing a hat, growing or removing a beard, or aging. The first organizations to make extensive use of facial recognition systems were the casinos, who used them to monitor players known to be cheaters. Airport immigration is a more recent, and rapidly growing application of the same technology.

While present-day facial recognition systems are nowhere near as reliable as they are depicted in movies and in television dramas—particularly in the case of recognizing a face in a crowd, which remains a difficult challenge—the technology is already useful in certain situations, and promises to increase in accuracy over the next few years.

The reason that facial recognition is of some use in casinos and airport immigration desks is that at those locations the target can be photographed alone, full face on, against a neutral background. But even then, there are difficulties. For example, in 2005, Germany started issuing biometric passports, but problems arose immediately due to people smiling. The German authorities had to issue guidelines warning that people "must have a neutral facial expression and look straight at the camera."

On the other hand, there are success stories. On December 25, 2004, the *Los Angeles Times* reported a police stop west of downtown Los Angeles, where police who were testing a new portable facial recognition system questioned a pair of suspects. One of the officers pointed the system, a hand-held computer with a camera attached, toward one of the two men. Facial recognition software in the device compared the image with those in a database that included photos of recent fugitives, as well as just over a hundred members of two notorious street gangs. Within seconds, the screen had displayed a gallery of nine faces with contours similar to the suspect's. The computer concluded that one of those images was the closest match, with a 94 percent probability of accuracy.

THE CASE OF THE SUSPICIOUS CONFERENCE CALLS

Detecting telephone fraud is another important application of neural networks.

Dr. Colleen McCue was, for many years, the program manager for the crime analysis unit at the Richmond Police Department in Richmond, Virginia, where she pioneered the use of data-mining techniques in law enforcement. In her book *Data Mining and Predictive Analysis*, she describes one particular project she worked on that illustrates the many steps that must often be gone through in order to extract useful information from the available data. In this case, a Kohonen neural net was used to identify clusters in the data, but as Dr. McCue explains, there were many other steps in the analysis, most of which had to be done by hand. Just as in regular police detective work, where far more time is spent on routine "slogging" and attention to details than on the more glamorous and exciting parts dramatized in movies and on TV, so too with data mining. Labor-intensive manipulation and preparation of the data by humans generally accounts for a higher percentage of the project time than the high-tech implementation of sophisticated mathematics. (This is, of course, not to imply that the mathematics is not important; indeed, it is often crucial. But much preparatory work usually needs to be done before the mathematics can be applied.)

The case McCue describes involves the establishment of a fraudulent telephone account that was used to conduct a series of international telephone conferences. The police investigation began when a telephone conference call service company sent them a thirty-seven-page conference call invoice that had gone unpaid. Many of the international conference calls listed on the invoice lasted for three hours or more. The conference call company had discovered that the information used to open the account was fraudulent. Their investigation led them to suspect that the conference calls had been used in the course of a criminal enterprise, but they had nothing concrete to go on to identify the perpetrators. McCue and her colleagues set to work to see if a data-mining analysis of the conference calls could provide clues to their identities.

The first step in the analysis was to obtain an electronic copy of the telephone bill in easily processed text format. With telephone records, this is fairly easy to do these days, but as data-mining experts the world over will attest, in many other kinds of cases a great deal of time and effort has to be expended at the outset in re-keying data as well as double-checking the keyed data against the hard-copy original.

The next stage was to remove from the invoice document all of the information not directly pertinent to the analysis, such as headers, information about payment procedures, and so forth. The resulting document included the conference call ID that the conference service issued for each call, the telephone numbers of the participants, and the dates and durations of the calls. Fewer than 5 percent of entries had a customer name, and although the analysts assumed those were fraudulent, they nevertheless kept them in case they turned out to be useful for additional linking.

The document was then formatted into a structured form amenable to statistical analysis. In particular, the area codes were separated from the other information, since they enabled linking based on area locations, and likewise the first three digits of the actual phone number were coded separately, since they too link to more specific location information. Dates were enhanced by adding in the days of the week, in case a pattern emerged.

At this point, the document contained 2,017 call entries. However, an initial visual check through the data showed that on several occasions a single individual had dialed in to a conference more than once. Often most of the calls were of short duration, less than a minute, with just one lasting much longer. The most likely explanation was that the individuals concerned had difficulty connecting to the conference or maintaining a connection. Accordingly, these duplications were removed. That left a total of 1,047 calls.

At this point, the data was submitted to a Kohonen-style neural network for analysis. The network revealed three clusters of similar calls, based on the day of the month that the call took place and the number of participants involved in a particular call.

Further analysis of the calls within the three clusters suggested the possibility that the shorter calls placed early in the month involved the leaders, and that the calls at the end of the month involved the whole group. Unfortunately for the police (and for the telephone company whose bill was not paid), at around that time the gang ceased their activity, so there was no opportunity to take the investigation any further. The analysts assumed that the sudden cessation was preplanned, since the gang organizers knew that when the bill went unpaid, the authorities would begin an investigation.

No arrests were made on that occasion. But the authorities did obtain a good picture of the conference call pattern associated with that kind of activity, and it is possible that, based on the findings of the study, the telephone company subsequently trained one of its own neural networks to look for similar patterns *as they occur*, to try to catch the perpetrators in the act. (This is the kind of thing that companies tend to keep secret, of course.)

Battles such as this never end. People with criminal intent will continue to look for ways to defraud the telecommunications companies. Data mining is the principal weapon the companies have in their arsenal to keep abreast of their adversaries.

MORE DATA MINING IN *NUMB3RS*

Given the widespread use of data-mining techniques in many areas of modern life, including crime detection and prevention, it is hardly surprising that Charlie mentions it in many episodes of *NUMB3RS*. For example, in the episode "Convergence," broadcast on November 11, 2005, a chain of robberies at upscale Los Angeles homes takes a more sinister turn when one of the homeowners is murdered. The robbers seem to have a considerable amount of inside information about the valuable items in the houses they rob and the detailed movements of the homeowners. Yet the target homes seem to have nothing in common, and certainly nothing that points to a source for the information the crooks are clearly getting. Charlie uses a data-mining program he wrote to look for patterns among all robberies in the area over the six-month period of the home burglaries, and eventually comes up with a series of car thefts that look as though they could be the work of the same gang, which leads to their capture.

Further Reading

Colleen McCue, *Data Mining and Predictive Analysis*, Butterworth-Heinemann (2007).

Jesus Mena, *Investigative Data Mining for Security and Criminal Detection*, Butterworth-Heinemann (2003).

CHAPTER

4 When Does the Writing First Appear on the Wall?

Changepoint Detection

THE BASEBALL NUMBERS GENIUS

In a third-season *NUMB3RS* episode entitled "Hardball," an aging baseball player, trying to make a comeback after several lackluster years in the minors, dies during on-field training. When the coach opens the dead player's locker, he finds a stash of needles and vials of steroids, and at once contacts the police. The coroner's investigation shows that the player suffered a brain hemorrhage resulting from a massive overdose of steroids, which he had started using to enhance his prospects of a return to the major league. But this was no accidental overdose. The drug in his locker was thirty times more powerful than the normal dosage, and had to have been prepared specially. The player had been murdered.

When Don is assigned to the case, he discovers some e-mails on the player's laptop from an unknown person who claimed to know that he was taking performance-enhancing drugs and threatened to inform the authorities. It looks like a case of blackmail. What is unusual is the proof that the unknown extortionist claimed to have. The e-mails have an attachment—a page of mathematical formulas that, the e-mailer claimed, showed exactly when in his professional career the player had started taking steroids.

Clearly, this was another case where Don would need the help of his younger brother. Charlie recognizes at once what the mathematics is about. "That's advanced statistical baseball analysis," he blurts out.

"Right, sabermetrics," replies Don, giving the accepted technical term for the use of statistics to analyze baseball performance.

The term "sabermetrics" is derived from the acronym SABR, which stands for the Society for American Baseball Research, and was coined by baseball statistics pioneer Bill James, one of the most enthusiastic proponents of using numbers to analyze the game.

Charlie also observes that whoever produced the formulas had devised his own mathematical abbreviations, something that might help identify him. Unfortunately, he does not know enough about the sabermetrics community to have any idea who might be behind the e-mail. But a colleague at CalSci has no trouble providing Charlie with the missing information. A quick search of several websites devoted to fantasy baseball soon reveals postings from an individual using the same mathematical notation.

For Don, the picture is now starting to emerge. The dead player had been killed to keep him from talking about the ring that was supplying him—and very likely other athletes—with illegal drugs. Obviously, the e-mails from the anonymous sabermetrician were what caused the fear that the narcotics ring would be discovered. But who was the killer: the e-mailer, the drug supplier, or someone else?

It does not take Don very long to trace the e-mail to a nerdy, twenty-five-year-old, high school dropout named Oswald Kittner, who used his self-taught mathematical abilities to make a fairly good living winning money by playing fantasy-league baseball. In this virtual arena, players create hypothetical teams of real players, which play against each other as computer simulations based on the current statistics for the real players. Kittner's success was based on his mathematical formulas, which turned out to be extremely good at identifying sudden changes in a player's performance—what is known in statistical circles as "changepoint detection."

As Charlie notes, what makes baseball particularly amenable to statistical analysis is the wealth of data it generates about individual performances coupled with the role of chance—e.g., the highly random result that comes with each pitch.

But Kittner had discovered that his math could do something else besides helping him to make a good living winning fantasy-league games. It could detect when a player started to use performance enhancing drugs. Through careful study of the performance and behavior of known steroid users in baseball, Kittner had determined the best stats to look for as an indication of steroid use—measuring long-ball hitting, aggressive play (being hit by pitches, for example), and even temper tantrums (arguments, ejections from games, and so forth). He had then created a mathematical surveillance system to monitor the best stats for all the players he was interested in, so that if any of them started using steroids, he would detect the changes in their stats and be able to react quickly. This would give him reliable information that a particular player is using steroids long before it becomes common knowledge.

"This is amazing," Charlie says as he looks again at the math. "This Kittner person has reinvented the Shiryayev–Roberts changepoint detection procedure!"

But was Kittner using his method to blackmail players or simply to win fantasy-league games by knowing in advance that a key player's performance was about to improve dramatically? Either way, before the young fan could put his new plan into action, one of his targets was murdered. And now the nerdy math whiz finds himself a murder suspect.

Kittner quickly comes clean and starts to cooperate with the authorities, and it does not take Don very long to solve the case.

CHANGEPOINT DETECTION

When it comes to crime, prevention is always better than trying to catch the perpetrators after the event. In some cases, the benefit of prevention can be much higher. For terrorist acts, such as those of September 11, 2001, the only way to preempt the attack is by getting information about the plotters before they can strike. This is what happened in the summer of 2006, when British authorities prevented a multiple attack on transatlantic planes using liquid explosives brought on board disguised as soft drinks and toiletries. A bioterrorist attack, on the other hand, may take weeks or months to reach full effect, as the pathogen works its way through the population. If the authorities can detect the pathogen in the relatively early stages of its dispersal, before its effect reaches epidemic proportions, it may be possible to contain it.

To this end, various agencies have instigated what is known as *syndromic surveillance*, where lists of pre-identified sets of symptoms are circulated among hospital emergency room personnel and certain other medical care providers, who must report to public health agencies if these symptoms are observed. Those agencies monitor such data continuously and use statistical analysis to determine when the frequency of certain sets of symptoms is sufficiently greater than normal to take certain predefined actions, including raising an alarm. Among the best-known systems currently in operation are RODS (Realtime Outbreak and Disease Surveillance) in Pennsylvania, ESSENCE (Early Notification of Community-Based Epidemics) in Washington, D.C., and the BioSense system implemented by the Centers for Disease Control and Prevention.

The principal challenge facing the designer of such a monitoring system is to identify when an activity pattern—say, a sudden increase in people taking time off from work because of sickness, or people visiting their doctor who display certain symptoms—indicates something unusual, above and beyond the normal ebb and flow of such activities. Statisticians refer to this task as *changepoint detection*—the determination that a definite change has occurred, as opposed to normal fluctuations.

In addition to syndromic surveillance—quickening the response to potential bioterrorist attacks by continuously collecting medical data, such as symptoms of patients showing up in emergency rooms—mathematical algorithms for changepoint detection are used to pinpoint other kinds of criminal and terrorist activity, such as

- Monitoring reports to detect increases in rates of certain crimes in certain areas
- Looking for changes in the pattern of financial transactions that could signal criminal activity

OUT OF INDUSTRY

The first significant use of changepoint detection systems was not for fighting crime, however, but for improving the quality of manufactured goods. In 1931, Walter A. Shewhart published a book explaining how to monitor manufacturing processes by keeping track of data in a control chart.

Shewhart, born in New Canton, Illinois, in 1891, studied physics at the Universities of Illinois and California, eventually earning a Ph.D., and was a university professor for a few years before going to work for the Western Electric Company, which made equipment for Bell Telephone. In the early days of telephones, equipment failure was a major problem, and everyone recognized that the key to success was to improve the manufacturing process. What Shewhart did was show how an ingenious use of statistics could help solve the problem.

His idea was to monitor an activity, such as a production line, and look for a change. The tricky part was to decide whether an unusual reading was just an anomaly—one of the random fluctuations that the world frequently throws our way—or else a sign that something had changed (a changepoint). (See Figure 4.)

Clearly, you have to look at some additional readings before you can know. But how many more readings? And how certain can you be that there really has been a change, and not just an unfortunate, but ultimately insignificant, run of unexpected readings? There is a trade-off to

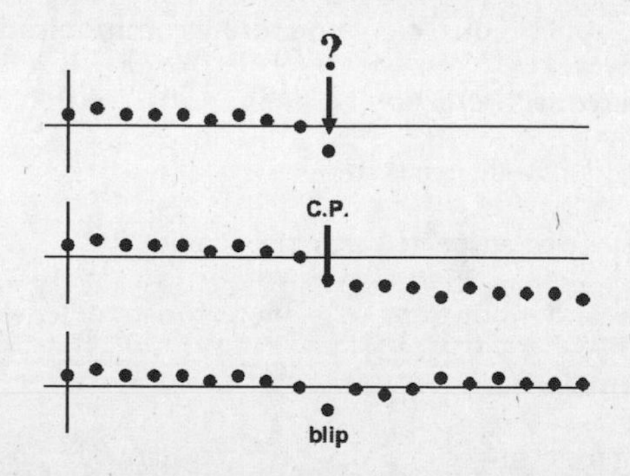

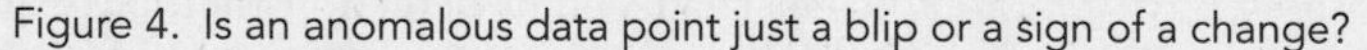
Figure 4. Is an anomalous data point just a blip or a sign of a change?

be made here. The more additional readings you take, the more confident you can be that there has been a change, but the longer you will have to wait before you can take action. Shewhart suggested a simple method that worked: You simply wait until you see an unusual result that is statistically well off the average, say three standard deviations. This method was a huge improvement, but it could still take a long time before a change was detected—too long for many applications, particularly those involved in crime detection and terrorism prevention. The key to a real advance was to use mathematics.

MATHEMATICS GETS INTO THE ACT

Around twenty-five years after Shewhart's book appeared, mathematicians in England (E. S. Page), the Soviet Union (A. N. Shiryayev), and the United States (S. W. Roberts) found several much more efficient (and mathematically sophisticated) ways to detect changepoints.

As the mathematical theory blossomed, so did the realization in industry and various branches of government (including law enforcement) that changepoint detection methods can be applied to a wide range of real-world problems. Such methods are now known to be useful in applications limited not only to industrial quality control but to such areas as:

- medical monitoring
- military applications (e.g., monitoring communication channels)
- environmental protection
- electronic surveillance systems
- surveillance of suspected criminal activity
- public health monitoring (e.g., bioterrorism defense)
- counterterrorism

To show how a more efficient changepoint detection method works, we'll focus on Page's procedure. (The Shiryayev–Roberts method that

Charlie Eppes mentions is slightly more technical to describe.) We'll look at an easier example than quality control: namely, detecting an increase in the frequency of some event.

Suppose that over some substantial period of time, it has been observed that a particular event occurs about once a month. Put another way, the probability of it happening on any given day is about 1 out of 30. Examples abound—a New Yorker finds a parking space on the street in front of her apartment, a husband actually *offers* to take out the garbage, a local TV news show doesn't lead off with a natural disaster or violent crime, and so on.

Now suppose that the frequency of a given event could increase dramatically—to once a week, say. We want to set up a changepoint detection system to react as quickly as possible without raising a false alarm too frequently.

The key issue we have to deal with is that chance fluctuations such as 3 or 4 occurrences in a single month can appear to indicate that the frequency has changed from once every 30 days to once every 7 days, even when there has not really been a change.

In the Page procedure, we introduce a numerical index, S, that tracks the activity. S is set initially equal to 1, and you revise S each day, using certain probability calculations, as we shall see shortly. When the value of S reaches or exceeds a certain pre-assigned level (we'll take 50 for the value in our example), you declare that a change has occurred. (Note that it is *not* required to estimate exactly *when* the change occurred, only to determine whether or not it *has* occurred.)

How do you "update" S each day? You multiply S by the probability of whatever happened that day, *assuming a change has already occurred*, and dividing it by the probability of whatever happened, *assuming a change has not occurred*.

For our example, this means that if the event occurs, you multiply S by $1/7$ and divide the result by $1/30$ (i.e., you multiply by 4.286); and if the event does not occur, you multiply S by $6/7$ and divide the result by $29/30$ (i.e., you multiply by 0.8867). In the former case, the value of S will increase. In the latter case, S decreases; if the new value of S is less than 1, you reset S to 1. (By never letting S be less than 1, the process remains in readiness to react to a change at any time.)

Because the event we're interested in is *more probable once a change has occurred*, on days when that event happens, S gets larger. And, not surprisingly, S gets smaller on days when the event does not happen.

This procedure is easy to carry out on a calculator. Suppose we start from scratch and see successive days as follows:

No, No, Yes (the event occurred), No, No, No, No, No, No, Yes, . . .

We start with S = 1. The first "No" gives S = 1 × .8867 = .8867, so we reset S = 1. The second "No" also gives S = .8867 and again we reset S = 1. Then we get a "Yes" and set S = 1 × 4.286 = 4.286. The following "No" gives S = 4.286 × .8867 = 3.800.

Continuing along the sequence of observations, we get the subsequent values 3.370, 2.988, 2.649, 2.349, 2.083, at which point we get the second "Yes", giving S = 8.927.

If we keep getting "Yes" this often, S will reach a threshold like 50 pretty quickly. But even *after* a change to 1 chance in 7 every day, it's not unusual to go two weeks without the event occurring, and that would multiply S by .8867 each day—unless the "never let S go below 1" rule kicks in.

If we use a computer to generate random days with a 1 out of 30 chance of the event every day, and each day is a new try, regardless of the history, it turns out that when a threshold of 50 is used for S, false indicators of a change will occur roughly 1,250 days apart—roughly three and a half years. Meanwhile, the quickness of detection after a change to 1 out of 7 chance every day, is on average no more than thirty-three days—about a month—even if the change occurs when S happens to be 1 (the lowest value possible), as at the beginning of the process. That's a lot better than Shewhart's procedure could do.

It turns out that the cost of getting a large interval between false change indicators (known to statisticians as the average run length, or ARL) in terms of increased time to detect a change, is not great with Page's procedure. Large increases in the ARL are accompanied by fairly small increases in detection time. Table 5 gives some results (for this example) illustrating the trade-off.

Threshold	ARL	Quickness of Detection
18.8	1.3 years	25.2 days
40	2.5 years	30.3 days
50	3.4 years	32.6 days
75	5.2 years	36.9 days
150	10.3 years	43.8 days

Table 5. The relationship between average run length and speed of detection.

So, for all that it is a great improvement on Shewhart's method, Page's procedure still seems to take a long time to reliably detect a change. Can we do better? Unfortunately, there are theoretical limitations to what can be achieved, as a mathematician named G. V. Moustakides proved in 1986. He showed that when the distributions of the data values before and after a possible change are known, as they are in our example, Page's procedure is the best you can do.

This fundamental limitation on the ability to reliably detect change-points is not merely frustrating to statisticians, it leaves society irrevocably vulnerable to threats in areas such as bioterrorism.

EARLY DETECTION OF A BIOTERRORIST ATTACK

A good example where changepoint detection is crucial is the *syndromic surveillance* we mentioned early in the chapter. The basic idea, which is being applied by many state and local health departments across the country, in cooperation with certain agencies of the federal government, goes like this: Suppose a terrorist attack uses an agent like anthrax or smallpox that can be released without causing an immediate alarm, so that the disease can spread for some time without alerting hospitals and public health officials.

In case of such an attack, it is critical for the authorities, particularly in the public health system, to be alerted as soon as possible so that they can figure out what is happening and take appropriate measures. These may include public warnings and bulletins to doctors and hospitals, describing what to look for in patients, how many people are likely to be affected and in which areas, and which methods to use in diagnosis and treatment.

Without having some system in place to accelerate the reaction of authorities, substantial delays could easily arise. Performing medical tests and confirming diagnoses can take time, and the possibility that the first patients may be few in number or scattered would contribute to the difficulty of recognizing a developing threat.

Faced with the limitations implied by Moustakides' 1986 result, researchers in the area of changepoint detection are constantly looking for better data sources to achieve the ultimate goal: the earliest possible detection of change.

In October 2006, the fifth annual Syndromic Surveillance Conference took place in Baltimore, Maryland. Research papers presented at the conference covered such topics as: Improving Detection Timeliness by Modeling and Correcting for Data Availability Delays; Syndromic Prediction Power: Comparing Covariates and Baselines; Efficient Large-scale Network-based Simulation of Disease Outbreaks; and Standard Operation Procedures for Three Syndromic Surveillance Systems in Washoe County, Nevada.

The greater the natural variability, the more severe is the problem of false alarms. But there is another aggravating factor: the sheer multiplicity of surveillance systems. The researchers at the conference pointed out that in the near future there may be thousands of such systems running simultaneously across the United States. Even if the frequency of false alarms is well controlled in each system, the overall rate of false alarms will be thousands of times greater, leading to obvious costs and concerns, including the classic "boy who cried wolf" phenomenon: Too many false alarms desensitize responders to real events.

How can the medical issues, the political issues, and the mathematical challenges associated with syndromic surveillance be addressed?

In several recent studies, researchers have used computer simulations to estimate how effective different mathematical methods will be in real-world performance. Results consistently show that when the Shewhart and Page approaches are compared, the latter is found to be superior. This is not a foregone conclusion, as the theorem of Moustakides, establishing that the Page procedure is the best possible, does not literally apply to the complicated problems that researchers are trying to solve. But mathematicians are used to the phenomenon that when a method or algorithm is proved to be the best possible in some simple situations it is likely to be close to the best one can do in more complicated situations.

Researchers are making intensive efforts to build a better foundation for success in using syndromic surveillance systems. The before-change scenarios require accurate knowledge of baseline data—that is, the appearance of patients in ERs with certain combinations of symptoms. The experts also pay considerable attention to the improvement of the probability estimates that go into the before-change part of the calculations. Several of the most common sets of symptoms that these surveillance systems look for have a greater probability of false positives during certain seasons of the year—cold and flu season, for example—so that the calculations are much more accurate when the baseline probabilities are defined in a way that reflects seasonal effects.

Another key to improving these systems is sharpening the probability estimates for after-change (post-attack) scenarios. One recent study examines the potential to improve biosurveillance by incorporating geographical information into the analysis. By building in statistical measures of the way the symptom reports cluster—in particular their spatial distribution as well as their temporal distribution—surveillance systems might gain greater power to detect outbreaks or abnormal patterns in disease incidence.

Mathematicians have other tricks up their sleeves that could help. The methods of Bayesian statistics (discussed in Chapter 6) can be used to incorporate certain kinds of useful information into changepoint detection calculations. Imagine that as we monitor a stream of data, looking for a changepoint, we have someone whispering hints in our ear—telling us at which points it is more likely or less likely that a change will occur. That is pretty much what the Department of Homeland Security's system of color-coded public alerts does, and the information gathered and assessed by intelligence agencies can be used to provide more focused alerts for certain types of disease-based terrorist attacks. Bayesian methods can incorporate such intelligence in a very natural and systematic way—in effect, lowering the threshold for raising an alarm during periods when the probabilities of particular kinds of bioterrorist attacks are heightened.

As one mathematician recently summed up the current situation in syndromic surveillance: "Changepoint detection is dead. Long live (even better) changepoint detection."

CHAPTER

5 Image Enhancement and Reconstruction

THE REGINALD DENNY BEATING

On April 29, 1992, at 5:39 PM, Reginald Oliver Denny, a thirty-nine-year-old, white truck driver loaded his red, eighteen-wheel construction truck with twenty-seven tons of sand and set off to deliver it to a plant in Inglewood, California. He had no idea that a little over an hour later, millions of television viewers would watch him being beaten to within an inch of his life by a rioting mob. Nor that the ensuing criminal prosecution of the rioters would involve a truly remarkable application of mathematics.

The sequence of events that led to Denny's beating had begun a year earlier, on March 3, 1991, when officers of the California Highway Patrol spotted a young black male, Rodney Glenn King, age twenty-six, speeding on Interstate 210. The officers chased King for eight miles at speeds in excess of 100 miles per hour, before finally managing to stop him in Lake View Terrace. When the CHP officers instructed him to lie down, King refused. At that point, a squad car of four Los Angeles Police Department officers arrived on the scene, and LAPD Sergeant Stacey Koon took command of the situation. When King then refused Sergeant Koon's command to comply with the instruction to lie down, Koon told his officers to use force. The police then started to hit King with their batons, and continued to beat him long after he had fallen to the ground. What the police did not know was that the entire event was being videotaped by a bystander, George Holliday, who would later sell the recording to the television networks.

Based largely on the videotapes, which were seen by television viewers all around the world, the four officers, three white and one Latino, were charged with "assault by force likely to produce great bodily injury" and with assault "under color of authority." As the officers' defense counsel argued in court, the video showed that King behaved wildly and violently throughout the incident (he was eventually charged with felony evasion, although that charge was later dropped), but as a result of the considerable attention given to Holliday's videotape, the focus was no longer on King but on the actions of the policemen. The court case unfolded against the volatile backdrop of a city where racial tensions ran high, and relations between the black community and the largely white LAPD were badly strained. When, on April 29, 1992, three of the officers were acquitted by a jury of ten whites, one Latino, and an Asian (the jury could not agree on a verdict for one of the counts on one of the officers), massive rioting erupted across the entire Los Angeles region.*

The riots would last for three days, making it one of the worst civil disturbances in Los Angeles history. By the time the police, Marine Corps, and National Guard restored order, there had been 58 riot-related deaths, 2,383 injuries, more than 7,000 fire responses, and damage to around 3,100 businesses amounted to over $1 billion. Smaller race riots broke out in other U.S. cities. On May 1, 1992, the third day of the Los Angeles riots, Rodney King went on television to appeal for calm and plead for peace, asking, "People, I just want to say, you know, can we all get along?"

But the rioting was just a few hours old as truck driver Reginald Denny turned off the Santa Monica Freeway and took a shortcut across Florence Avenue. At 6:46 PM, after entering the intersection at Normandie, he found himself surrounded by black rioters who started to throw rocks at his windows, and he heard people shouting at him to stop. Overhead, a news helicopter piloted by reporter Bob Tur captured the events that followed.

---

*After the riots, federal charges of civil rights violations were brought against the four officers. Sergeant Stacey Koon and Officer Laurence Powell were found guilty; the other two were acquitted.

One man opened the truck door, and others dragged Denny out. Denny was knocked to the ground and one of the assailants held his head down with his foot. Denny, who had done nothing to provoke the violence, was kicked in the stomach. Someone hurled a five-pound piece of medical equipment at Denny's head and hit him three times with a claw hammer. Still another man threw a slab of concrete at Denny's head and knocked him unconscious. The man, who would subsequently be identified as Damian Williams, then did a victory dance over Denny, flashing a gang sign at the news helicopter hovering above, which was broadcasting the events on live television, and pointed at Denny. Another rioter then spat on Denny and left with Williams. Several passersby took pictures of the attack but no one came to Denny's aid.

After the beating ended, various men threw beer bottles at the unconscious Denny. Someone came along and riffled through Denny's pockets, taking his wallet. Another man stopped near the body and attempted to shoot the gas tank of Denny's truck but missed. Eventually, with the attackers gone, four men who had been watching the events on TV came to Denny's aid. One of them was a trucker with a license that allowed him to drive Denny's truck. The four rescuers loaded the prostrate trucker into his cab and drove him to the hospital. Upon arrival at the hospital, Denny suffered a seizure.

Paramedics who attended to Denny said he came very close to death. His skull was fractured in ninety-one places and pushed into the brain. His left eye was so badly dislocated that it would have fallen into his sinus cavity had the surgeons not replaced the crushed bone with a piece of plastic. A permanent crater remains in his head to this day, despite efforts to correct it.

Based on identification from the TV news video taken from Bob Tur's helicopter, the three men most directly involved in the attack on Denny were arrested and brought to trial. Of the three, only one, Damian Williams, would be convicted, and then only on one felony charge, the court seeming to take the view (rightly or wrongly) that the acts were not premeditated and were the result of citywide mob mentality. For our present purpose, however, the most fascinating aspect of the case is that the identification of Williams was a result of some remarkable new mathematics, and the acceptance of those methods by the court was a milestone in legal history.

THE ROSE TATTOO

Although millions watched the attack on Denny on TV, either live or during endless repeats on news programs, and although the prosecution in the trial of Williams and his two accused accomplices showed forty minutes of video recordings of the event as evidence, identification of the assailants of sufficient reliability to secure a conviction proved difficult. The video footage had been shot from a small portable camera, handheld by Tur's wife, Marika, in a helicopter hovering above the scene. The result was grainy and blurred, and on no occasion did Marika Tur get a clear face shot of the assailants. The person shown throwing a large slab of concrete at Denny's head and then performing a victory dance over the victim's now unconscious body *could* have been Williams. But it equally could have been any one of hundreds of young black males in the Los Angeles area who shared his overall build and appearance.

One feature that did distinguish Williams from other possible suspects was a large tattoo of a rose on his left arm. (The tattoo identified him as a member of the notorious Los Angeles gang Eight Tray Gangster Crips.) Unfortunately, although some frames of the newsreel video did show the assailant's left arm, the image was not sharp enough to discern the tattoo.

At that point, the frustrated prosecutors got a major break. A Santa Monica reporter supplied them with some still photographs shot from a helicopter with a 400-millimeter long-distance lens. Thanks to the much higher resolution of still photographs, close scrutiny of one of the photographs, both with the naked eye and a magnifying glass, did reveal a vague gray region on the assailant's left arm as he stood over the prone body of Williams. (See Figure 5.) The gray region—a mere one six-thousandth of the overall area of the photograph—might indeed have been a tattoo; unfortunately, it could just as easily have been a smudge of dirt or even a blemish on the photo. Enter mathematics.

Using highly sophisticated mathematical techniques, developed initially to enhance surveillance photographs taken by military satellites, the crucial

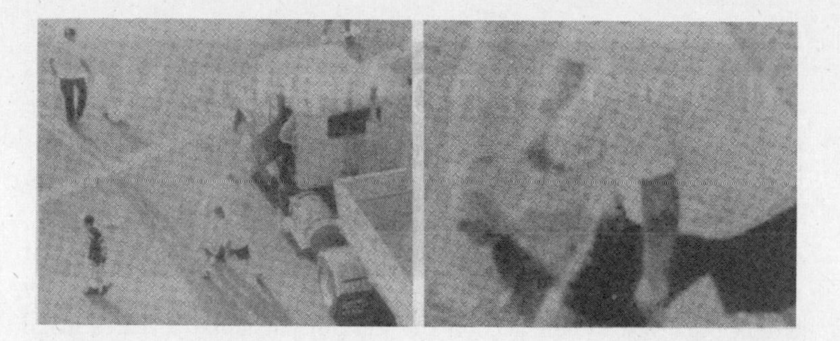

Figure 5. Mathematically enhanced aerial photograph of the Reginald Denny beating, with feature enlargement showing a blurred mark on the assailant's left arm.

portion of the photograph was processed on a high-performance computer to generate a much clearer image. The resulting image revealed that the apparent mark on the suspect's left arm had a shape and color that, above the usual legal threshold of "beyond a reasonable doubt," was indeed a rose tattoo like the one on Damian Williams' arm.

The techniques used to process the photographic images in the Reginald Denny case fall in the general area known as *image enhancement*. This is not a technique for adjusting brightness, color, or contrast, or otherwise tweaking photographs familiar to computer users in the form of programs such as Photoshop, nor is it the proprietary photograph-handling software that often comes with new digital cameras. In image enhancement, mathematical techniques are used to *reconstruct* image details that were degraded by optical blurring in the original photograph.

The term "reconstruct" as used here can be misleading to laypersons unfamiliar with the technique. One of the key steps in the trial of Damian Williams was for the experts to convince the judge, and then the jury, that the process was reliable, and that the resulting image did not show "what *might* have been," but did in fact reveal "what *was*." The judge's ruling in the case, that images produced by enhancement techniques were indeed allowable evidence, was a landmark in legal history.

The general idea behind image enhancement is to use mathematics to supply features of the image that were not captured in the original photograph. No photograph will represent everything in a visual scene. Most photographs capture enough information that the human eye is often unable to discern any differences between the photograph and the original scene, and certainly enough for us to identify an individual. But as cognitive scientists have demonstrated, much of what we see when we look at either a real-life scene or a photograph is supplied by our brains, which fill in—generally reliably and accurately—anything that (for one reason or another) is missing from the visual signal that actually enters our eyes. When it comes to certain particular features in an image, mathematics is far more powerful, and can furnish—also reliably and accurately—details that the photograph never fully captured in the first place.

In the Damian Williams trial, the key prosecution witness who identified the defendant was Dr. Leonid Rudin, the cofounder in 1988 of Cognitech, Inc., a Santa Monica–based company specializing in image processing. As a doctoral student at Caltech in the mid-1980s, Rudin developed a novel method for deblurring photographic images. Working with his colleagues at Cognitech, Rudin further developed the approach to the point where, when the Williams trial came to court, the Cognitech team was able to take video images of the beating and process them mathematically to produce a still image that showed what in the original video looked like a barely discernible smudge on the forearm of one of the assailants to be clearly identifiable as a rose tattoo like the one on Williams' arm. When the reconstructed photograph was presented to the jury for identification, Williams' defense team at once changed its position from "Williams is not the person in the photo/video" to his being a "nonpremeditated" participant in the attack.

WHAT THE EYE CANNOT SEE: THE MATH OF IMAGE RECONSTRUCTION

To get some idea of the kind of problem facing the Cognitech engineers, imagine that we are faced with the comparably simpler task of simply enlarging a photograph (or part of a photograph) to twice its original size. (Enlargement of the key part of the Williams image was in fact one of the things Rudin and his colleagues did as part of their analysis.) The simplest approach is to add more pixels according to some simple rule. For example, suppose you start with an image stored as a 650 × 500 pixel grid and want to generate an enlarged version measuring 1300 × 1000 pixels. Your first step is to double the dimensions of the image by coloring the pixel location (2x,2y) the same as location (x,y) in the original image. This generates an image twice as large, but having lots of "holes" and hence being very grainy. (None of the pixels with at least one odd coordinate has a color.) To eliminate the graininess you could then color the remaining locations (the ones having at least one odd coordinate) by taking the mean of the color values for all adjacent pixels in the evens-evens grid.

Such a naïve method of filling in the holes would work fine for fairly homogeneous regions of the image, where changes from one pixel to the next are small, but where there is an edge or a sudden change in color, it could be disastrous, leading to, at best, blurred edges and, at worst, significant distortion (pixelation) of the image. Where there is an edge, for instance, you should really carry out the averaging procedure along the edge (to preserve the geometry of the edge) and then average separately in the two regions on either side. For an image with just a few, well-defined, and essentially straight edges, you could set this up by hand, but for a more typical image you would want the edge detection to be done automatically. This requires that the image-processing software can recognize edges. In effect, the computer must be programmed with the capacity to "understand" some features of the image. This can be done, but it is not easy, and requires some sophisticated mathematics.

The key technique is called segmentation—splitting up the image into distinct regions that correspond to distinct objects or parts of objects in the original scene. (One particular instance of segmentation is distinguishing objects from the background.) Once the image has been segmented, missing information within any given segment can be re-introduced by an appropriate averaging technique. There are several different methods for segmenting an image, all of them very technical, but we can describe the general idea.

Since digital images are displayed as rectangular arrays of pixels, with each pixel having a unique pair of x,y coordinates, any smooth edge or line in the image may be viewed as a curve, defined by an algebraic in the classical sense of geometry. For example, for a straight line, the pixels would satisfy an equation of the form

$$y = mx + c$$

Thus, one way to identify any straight-line edges in an image would be to look for collections of pixels of the same color that satisfy such an equation, where the pixels to one side of the line have the same color but the pixels on the other side do not. Likewise, edges that are curved could be captured by more complicated equations such as polynomials. Of course, with a digitized image, as with a scene in the real world, the agreement with an equation would not be exact, and so you'd have to allow for a reasonable amount of approximation in satisfying the equation. Once you do that, however, then you can take advantage of a mathematical fact that any smooth edge (i.e., one that has no breaks of sharp corners) can be approximated to whatever degree of accuracy you desire by a collection of (different) polynomials, with one polynomial approximating one part of the edge, another polynomial approximating the next part of the edge, and so on. This process will also be able to handle edges having sharp corners; at a corner, one polynomial takes over from the previous one.

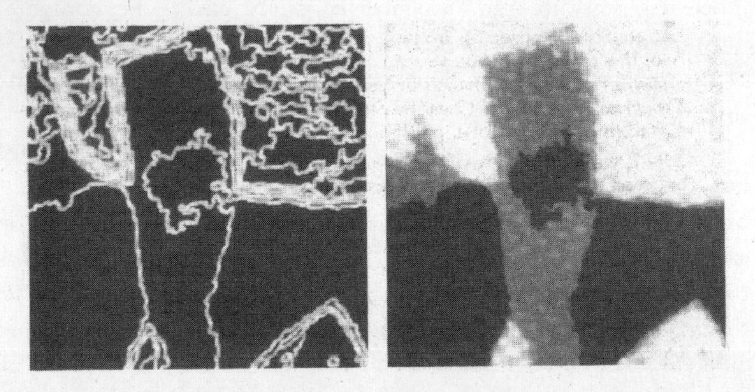

Figure 6. The result of the segmentation algorithm run on the photograph of the left arm of the Reginald Denny assailant, showing a mark entirely consistent with the rose tattoo on Damian Williams' left arm.

This simple idea shows how the problem of verifying that a given edge *is indeed an edge* can be reduced to a problem about finding equations. Unfortunately, being able to find an equation whose curve approximates a segment of a *given* edge does not help you identify that edge in the first place. For humans, recognizing an edge is generally not a problem. We (and other creatures) have sophisticated cognitive abilities to recognize visual patterns. Computers, however, do not. What they do excel at is manipulating numbers and equations. Thus, the most promising approach to edge detection would seem to be to manipulate equations in some systematic way until you find one that approximates the given edge segment—that is, such that the coordinates of the points on the edge segment approximately satisfy the equation. Figure 6 shows the result of the Cognitech segmentation algorithm applied to the crucial left-arm portion of the aerial photograph in the Reginald Denny beating case.

This, in essence, is how segmentation works, but the successful implementation requires the use of mathematics well beyond the scope of this book. For the benefit of readers with some familiarity with college-level mathematics, the following section provides a brief account of the method; readers without the requisite background should simply skip the section.

IMAGE ENHANCEMENT: THE INSIDE SCOOP

Image enhancement is easier with black-and-white (more precisely, gray-scale) images than full color, so we'll look just at that special case. Given this restriction, a digital image is simply a function F from a given rectangular space (say, a grid 1000 × 650) into the real unit interval [0,1] (i.e., the real numbers between 0 and 1 inclusive). If F(x,y) = 0, then pixel (x,y) is colored white, if F(x,y) = 1, the pixel is colored black, and for all other cases F(x,y) denotes a shade of gray between white and black; the greater the value of F(x,y), the closer the pixel (x,y) is to being black. In practice, a digital image assigns gray-scale values to only a finite number of pixels—the image consists of a grid of pixels. To do the mathematics, however, we assume that the function F(x,y) is defined on the entire rectangle, that is to say, F(x,y) gives a value for any real numbers x, y within the stipulated rectangle. This allows us to use the extensive and powerful machinery of two-dimensional calculus (i.e., calculus of real-valued functions of two real variables).

The method used by the Cognitech team was based on an idea Rudin had when working as a graduate student intern at Bell Laboratories in the early 1980s, and developed further in his doctoral dissertation submitted at Caltech in 1987. By asking himself fundamental questions about vision—"Why do we see a single point on a sheet of paper?" or "How do we see edges?" or "Why is a checker board pattern with lots of squares so annoying to the eye?" or "Why do we have difficulty understanding blurry images?"—and linking those questions to the corresponding mathematical function F(x,y), he realized the significance of what are called the singularities in the function. These are the places where the derivative (in the sense of calculus) becomes infinite. This led him to focus his attention on a particular way to measure how close a particular function is to a given image—the so-called total variation norm. The details are highly technical, but not required here. The upshot was that, together with colleagues at Cognitech, Rudin developed computational techniques to restore images using what is now called the total variation method.*

MATH IN COURT

In addition to its obvious uses in military intelligence, the methods Cognitech developed found early applications in the enhancement of satellite images for nonmilitary purposes such as oil spill detection, and in the processing of images obtained by MRIs to identify tissue abnormalities such as tumors or obstructed arteries. By the time of the Damian Williams trial, the company was well established and ideally placed to make its groundbreaking contribution.

In addition to enhancing the key image that identified Damian Williams as the man who threw a concrete slab at Denny's head, Rudin and his colleagues also used their mathematical techniques to obtain other photograph-quality still images from video footage of the events,

---

*For those who know the lingo, the key idea is to use Euler-Lagrange PDE minimization, a calculus technique developed long before computers came onto the scene, on the total variation functional.

thereby identifying Williams as the perpetrator of assaults on several other victims as he moved from place to place that day.

Anyone who has viewed a freeze-frame of a video recording on a VCR will have observed that the quality of the image is extremely low. Video systems designed for home or even for news reporting take advantage of the way the human visual system works, to reduce the camera's storage requirements. Roughly speaking, each frame records just half the information captured by the lens, with the next frame recording (the updated version of) the missing half. Our visual system automatically merges the two successive images to create a realistic-looking image as it perceives the entire sequence of still images as depicting continuous motion. Recording only half of each still image works fine when the resulting video recording is played back, but each individual frame is usually extremely blurred. The image could be improved by merging together two successive frames, but because video records at a much lower resolution (that is, fewer pixels) than a typical still photograph, the result would still be of poor quality. To obtain the photograph-quality images admissible in court as evidence, Rudin and his Cognitech team used mathematical techniques to merge not two but multiple frames. Mathematical techniques were required because the different frames captured the action at different times; simply "adding together" all of the frames would produce an image even more blurred than any single frame.

The sequence of merged still images produced from the videotapes seemed to show Williams committing a number of violent acts, but identification was not always definitive, and as the defense pointed out, the enhanced images seemingly raised some issues. In one case, later images showed a handprint on the perpetrator's white T-shirt that was not visible on earlier images. This was resolved when a close examination of the videotape indicated the exact moment the handprint was made. More puzzling, earlier images showed a stain on the attacker's T-shirt that could not be seen on later images. On that occasion, targeted image enlargement and enhancement showed that in the later shots, the perpetrator was wearing two white T-shirts, one over the other, with the outer one hiding the stain on the inner one. (The enhanced image revealed the band around the bottom of the inner T-shirt protruding below the bottom of the outer shirt.)

Cognitech's video-processing technology also played a role in some of the other legal cases resulting from the riots. In one of them, the defendant, Gary Williams, pleaded guilty to all charges after the introduction into court of an enhanced ninety-second videotape that showed him rifling Denny's pockets and performing other illegal acts. Although Gary Williams had intended to plead not guilty and go for a jury trial, when he and his counsel saw the enhanced video, they acknowledged that it was sufficiently clear that a jury might well accept it as adequate identification, and decided instead to go for a plea bargain, resulting in a three-year sentence.

AND ON INTO THE FUTURE

With the L.A. riots cases establishing the legal admissibility of enhanced images, it was only a few weeks before Cognitech was once again asked to provide its services. On that occasion, they were brought in by the defense in a case involving an armed robbery and shooting at a jewelry store. The robbery had been captured by a surveillance camera. However, not only was the resolution low (as is often the case), the camera recorded at a low rate of one frame per second, well below the threshold to count as true video (roughly twenty-four frames per second). Rudin and his colleagues were able to construct images that contradicted certain testimony presented at trial. In particular, the images obtained showed that one key witness was in a room where she could not possibly have seen what she claimed to have seen.

Since then, Cognitech has continued to develop its systems, and its mathematical state-of-the-art Video-Investigator and Video-Active Forensic Imaging software suite is in use today by thousands of law enforcement and security professionals, and in forensic labs throughout the world, including the FBI, the DEA, the U.K. Home Office and Scotland Yard, Interpol, and many others.

In one notable case, a young African-American adult in Illinois was convicted (in part based on his own words, in part on the basis of videotape evidence) of the brutal murder of a store clerk, and was facing the death penalty. The accused and his family were too poor to seek costly expert services, but by good fortune his public defender questioned the videotape-based identification made by the state and federal forensic experts. The defender contacted Cognitech, which carried out a careful video restoration followed by a 3-D photogrammetry (the science of accurate measuring from photographs, using the mathematics of 3-D perspective geometry) of the restored image. This revealed an uncontestable discrepancy with the physical measurements of the accused. As a result, the case was dismissed and the innocent young man was freed. Some time later, an FBI investigation resulted in the capture and conviction of the real murderer.

Working with the Discovery Channel on a special about UFO sightings in Arizona (*Lights over Phoenix*), Cognitech processed and examined video footage to show that mysterious "lights" seen in the night sky were consistent with light flares used by the U.S. Air Force that night. Furthermore, the Cognitech study revealed that the source of the lights was actually behind the mountains, not above Phoenix as observers first thought.

Most recently, working on another Discovery Channel special (*Magic Bullet*) about the J.F.K. assassination, Rudin and his team used their techniques to solve the famous grassy knoll "second shooter" mystery. By processing the historic Mary Moorman photo with the most advanced image-restoration techniques available today, they were able to show that the phantom "second shooter" was an artifact of the photograph, not a stable image feature. Using advanced 3-D photogrammetric estimation techniques, they measured the phantom "second shooter" and found it to be three feet tall.

In an era when anyone with sufficient skill can "doctor" a photograph (a process that also depends on sophisticated mathematics), the old adage "photographs don't lie" no longer holds. But because of the development of image-reconstruction techniques, a new adage applies: Photographs (and videos) can generally tell you much more than you think.

CHAPTER

6 Predicting the Future
Bayesian Inference

MANHUNT

When a bus transporting prison inmates is involved in a road accident, two of the prisoners escape, killing the guard in the process. Charlie provides some help in unraveling the case by carrying out a detailed analysis of the crash scene, which enables him to reconstruct what must have happened. His conclusion: The crash was not an accident, it was staged. The escape was planned.

This is the story *NUMB3RS* viewers watched in the first-season episode called "Manhunt," broadcast on May 13, 2005.

Charlie's fictional mathematical reconstruction of the accident is based on the way accident investigators work in real life. But figuring out how the crash occurred is not the end of Charlie's involvement in this particular case. After one of the escapees is captured, attention focuses on finding the other, the man who planned the escape. The recaptured prisoner, a model prisoner who had almost completed his sentence, turns out to have had no prior knowledge of the escape plot. But he is able to tell Don about his companion, a convicted killer serving a life sentence with no possibility of parole—and hence a highly dangerous individual with little to lose from killing again. The most chilling thing the recaptured prisoner tells Don is that the killer intends to murder the key witness at his trial, a woman whose testimony had helped convict him.

Don tries to persuade the witness to leave town and go into hiding until the killer is caught, but she refuses. She is a hospital doctor with patients she feels she cannot walk away from. This places Don in a race against the clock to track down the escapee before he can make good his deadly intention.

Media coverage of the escape, including police photographs of the escaped killer, soon leads to reports of sightings from members of the public. Unfortunately, the reports flood in, several hundred in all, and they are scattered across Los Angeles, many of them claiming simultaneous sightings at locations miles apart. While some of the reports may be hoaxes, most are probably from well-meaning citizens who genuinely believe they have spotted the man whose photograph they had seen in the newspaper or on TV. But how can Don decide which sightings are accurate—or even which ones are most likely to be correct?

This is where Charlie makes his second contribution to the case. He says he has carried out a "Bayesian statistical analysis" of the sightings, which tells him which particular sightings are most likely reliable. Using Charlie's results, Don is able to determine where the killer probably is, and manages to get to him just in time to prevent him from killing the witness.

As is often the case with dramatic portrayals of mathematics or science at work, the length of time available to Charlie to produce his ranking of the reported sightings is significantly shortened, but the idea of using the mathematically based technique known as Bayesian analysis is sound. At the end of this chapter, we'll explain how Charlie most likely performed his analysis. (Viewers do not see him carrying out this step, and the script offers no details.) First, though, we need to describe in more general terms the hugely important techniques of Bayesian statistics.

PREDICTING THE FUTURE

Law enforcement would be much easier if we could look into the future and know about crimes before they actually occur.* Even with the help of mathematics, however, this is not possible. Mathematics can predict with as much accuracy as you wish the position of a spacecraft traveling at thousands of miles an hour at noon Greenwich mean time six months

---

*This was the main plot idea behind the 2002 blockbuster movie *Minority Report*, starring Tom Cruise. But that, of course, is fiction.

from now, but most of us find it hard to predict with any accuracy where we will be at noon even a week from now. Human behavior simply is not amenable to mathematical prediction. At least, not if you want the mathematics to give an exact answer. If, however, you are willing to settle for numerical estimates on things *likely* to happen, then mathematics can be of real use.

For instance, no one apart from the handful of Al Qaeda operatives who carried out the September 11, 2001, attacks knew in advance what was going to take place. But things might have turned out very differently if the U.S. authorities had known that such an attack was likely, what the most probable targets were, and which actions to take to prevent the terrorists from carrying out their plan. Could mathematics help provide such advance warning of things that might occur, perhaps with some kind of numerical measure of their likelihood?

The answer is, not only is this possible, it actually happened. A year before the attack took place, mathematics had predicted that the Pentagon was a likely terrorist target. On that occasion, no one took the mathematical prediction sufficiently seriously to do something about it. Of course, it's always easier to be smart after the event. What mathematics can do—and did—is (as we explain below) furnish a list of likely targets, together with estimates of the probabilities that an attack will take place. Policymakers still have to decide which of the many threats identified should be singled out for expenditure of the limited resources available. Still, given how events unfolded on that fateful day in 2001, perhaps next time things will turn out differently.

HOW MATHEMATICS PREDICTED THE 9/11 ATTACK ON THE PENTAGON

In May 2001, a software system called Site Profiler was fielded to all U.S. military installations around the world. The software provided site commanders with tools to help to assess terrorist risks, to manage those risks, and to develop standardized antiterrorism plans. The system worked by combining different data sources to draw inferences about the risk of terrorism, using a mathematical technique called Bayesian inference.

Prior to the system's deployment, its developers carried out a number of simulation tests, which they referred to in a paper they wrote the year before.* Summarizing the results of the tests, they noted: "While these scenarios showed that the RIN [Risk Influence Network] 'worked,' they tended to be exceptional (e.g., attacks against the Pentagon)."

As the world now knows, the Pentagon was the site of an attack. Unfortunately, neither the military command nor the U.S. government had taken seriously Site Profiler's prediction that the Pentagon was in danger—nor, for that matter, had the system developers themselves, who viewed the prediction as "exceptional."

As experience has taught us time and time again, human beings are good at assessing certain kinds of risks—broadly speaking, personal risks associated with familiar situations—but notoriously bad at assessing others, particularly risks of novel kinds of events. Mathematics does not have such a weak spot. The mathematical rules the developers built into Site Profiler did not have an innate "incredulity factor." Site Profiler simply ground through the numbers, assigning numerical risks to various events, and reported the ones that the math said were most probable. When the numbers said the Pentagon was at risk, that's what the program reported. Humans were the ones who discounted the prediction as too far-fetched.

This story tells us two things. First, that mathematics provides a powerful tool for assessing terrorist risks. Second, that humans need to think very carefully before discounting the results that the math produces, no matter how wild they might seem.

This is the story behind that math.

SITE PROFILER

Site Profiler was licensed by the U.S. Department of Defense in 1999 to develop an enterprise-wide antiterrorism risk management system called the Joint Vulnerability Assessment Tool (JVAT).

**An Application of Bayesian Networks to Antiterrorism Risk Management for Military Planners,* by Linwood D. Hudson, Bryan S. Ware, Suzanne M. Mahoney, and Kathryn Blackmond Laskey.

The JVAT program was started in response to the bombing of U.S. Air Force servicemen in Khobar Towers, Saudi Arabia, in June 1996, in which nineteen American servicemen and one Saudi were killed and 372 of many nationalities wounded, and the August 1998 bombings of United States embassies in the East African capital cities of Dar es Salaam, Tanzania, and Nairobi, Kenya, where a total of 257 people were killed and more than 4,000 wounded.

The investigations into those events revealed that the United States had inadequate methods for assessing terrorist risks and anticipating future terrorist events. Addressing that need was a major challenge. Since the intentions, methods, and capabilities of potential terrorists, and often even their identities, can almost never be forecast with certainty from the intelligence information available, much of the effort in countering the threat has to focus on identifying likely targets. Understanding the vulnerabilities of a potential target and knowing how to guard against attacks typically requires input from a variety of experts: physical security experts, engineers, scientists, and military planners. Although a limited number of experts may be able to understand and manage one or two given risks, no human can manage all of the components of hundreds of risks simultaneously. The solution is to use mathematical methods implemented on computers.

Site Profiler is just one of many systems that allow users to estimate—with some degree of precision—and manage a large risk portfolio by using Bayesian inference (implemented in the form of a Bayesian network, which we describe below) to combine evidence from different data sources: analytic models, simulations, historical data, and user judgments.

Typically, the user of such a system (often an expert assessment team) enters information about, say, a military installation's assets through a question-and-answer interface reminiscent of a tax preparation package. (Site Profiler actually modeled its interface on Turbo Tax.) The software uses the information it has gathered to construct mathematical objects to represent the installation's various assets and threats, to express the entire situation as a Bayesian network, to use the network to evaluate the various risks, and finally to output a list of threats, each one given a numerical rank based on its likelihood, the severity of its consequences, and so forth. Our interest here is in the mathematics that sits "under the hood" of such a system.

The key idea behind all this goes back to an eighteenth-century English clergyman, Thomas Bayes.

THOMAS BAYES AND THE PROBABILITIES OF WHAT WE KNOW

In addition to being a Presbyterian minister, Thomas Bayes (1702–1761) was a keen amateur mathematician. He was fascinated by how we come to know the things we know, specifically how we judge the reliability of information we acquire, and he wondered whether mathematics could be used to make such judgments more precise and accurate. His method of calculating how our beliefs about probabilities should be modified whenever we get new information—new *data*—led to the development of Bayesian statistics, an approach to the theory and practice of statistical analysis that has long attracted passionate adherents, as well as staunch critics. With the advent in the late twentieth century of immensely powerful computers that can crunch millions of pieces of data per second, both Bayesian statisticians (who *always* use his fundamental idea) and non-Bayesian statisticians (who *sometimes* use it) owe him a great debt.

BAYES' METHOD

Bayes' idea concerns probabilities about things that may or may not be true—that the probability of heads in a coin flip is between .49 and .51; that Brand Y cures headaches more frequently than Brand X; that a terrorist or criminal will attack target J or K or L. If we want to compare two possibilities, say, A and B, Bayes gives the following recipe:

1. Estimate their relative probabilities P(A) / P(B)—the odds of A versus B.

2. For each observation of new information, X, calculate the likelihood of that observation if A is true and if B is true.

3. Re-estimate the relative probabilities of A and B as follows: P(A given X) / P(B given X) = P(A)/P(B) × Likelihood Ratio, where the Likelihood Ratio is the likelihood of observing X if A is true divided by the likelihood of observing X if B is true.
4. Repeat the process whenever new information is observed.

The odds of A versus B in step one are called "prior odds," meaning that they represent our state of knowledge prior to observing the data X. Often this knowledge is based on subjective judgments—say, what are the odds that a new drug is better than the standard drug for some illness, or what are the odds that terrorists will attack one target versus another, or perhaps even what are the odds that a criminal defendant is guilty before any evidence is presented? (The arbitrariness of putting a number on the last example is one reason that the use of Bayesian statistics in criminal trials is essentially zero!)

To understand Bayes' recipe, it is helpful to consider an example where these "prior odds" are actually known. When that situation occurs, the use of Bayesian methods is noncontroversial.

THE (FICTITIOUS) CASE OF THE HIT-AND-RUN ACCIDENT

A certain town has two taxi companies, Blue Cabs and Black Cabs. Blue Cabs has 15 taxis, Black Cabs has 75. Late one night, there is a hit-and-run accident involving a taxi. The town's 90 taxis were all on the streets at the time of the accident. A witness sees the accident and claims that a blue taxi was involved. At the request of the police, the witness undergoes a vision test under conditions similar to the those on the night in question. Presented repeatedly with a blue taxi and a black taxi, in random order, he shows he can successfully identify the color of the taxi 4 times out of 5. (The remaining one fifth of the time, he misidentifies a blue taxi as black or a black taxi as blue.) If you were investigating the case, which company would you think is most likely to have been involved in the accident?

Faced with eyewitness evidence from a witness who has demonstrated that he is right 4 times out of 5, you might be inclined to think it was a blue taxi that the witness saw. You might even think that the odds in favor of it being a blue taxi were exactly 4 out of 5 (that is, a probability of 0.8), those being the odds in favor of the witness being correct on any one occasion.

Bayes' method shows that the facts are quite different. Based on the data supplied, the probability that the accident was caused by a blue taxi is only 4 out of 9, or 44 percent. That's right, the probability is less than half. It was more likely to have been a black taxi. Heaven help the owner of the blue taxi company if the jurors can't follow Bayesian reasoning!

What human intuition often ignores, but what Bayes' rule takes proper account of, is the 5 to 1 odds that any particular taxi in this town is black. Bayes' calculation proceeds as follows:

1. The "prior odds" of a taxi being black are 5 to 1 (75 black taxis versus 15 blue).
   The likelihood of X="the witness identifies the taxi as blue" is:
   1 out of 5 (20%) if it is black
   4 out of 5 (80%) if it is blue.

2. The recalculation of the odds of black versus blue goes like this:
   P(taxi was black given witness ID)/ P(taxi was blue given witness ID) =
   $(5 / 1) \times (20\% / 80\%) = (5 \times 20\%) / (1 \times 80\%) = 1 / .8 = 5/4.$

Thus Bayes' calculation indicates that the odds are 5 to 4 after the witness' testimony that the taxi was black.

If this seems counterintuitive (as it does initially to some people) consider the following "thought experiment." Send out each of the 90 taxis on successive nights and ask the witness to identify the color of each under the same conditions as before. When the 15 blue taxis are seen, 80% of the time they are described as blue, so we can expect 12 "blue sightings" and 3 "black sightings." When the 75 black taxis go out, 20% of the time they are described as blue, so we can expect 15 "blue sightings" and 60 "black sightings." Overall, we can expect 27 taxis will be described by the witness as "blue", whereas only 12 of them actually were blue and 15 were black. The ratio of 12 to 15 is the same as 4 to 5—in other words, only 4 times out of every 9 (44 percent of the time) when the witness says he saw a blue taxi was the taxi really blue.

In an artificial scenario where the initial estimates are entirely accurate, a Bayesian network will give you an accurate answer. In a more typical real-life situation, you don't have exact figures for the prior probabilities, but as long as your initial estimates are reasonably good, then the method will take account of the available evidence to give you a *better* estimate of the probability that the event of interest will occur. Thus, in the hands of an expert, someone who is able to assess all the available evidence reliably, Bayesian networks can be a powerful tool.

HOW CHARLIE HELPED TRACK DOWN THE ESCAPED KILLER

As we mentioned at the start of the chapter, nothing in the "Manhunt" episode of *NUMB3RS* explained how Charlie analyzed the many reported sightings of the escaped convict. Apart from saying that he used "Bayesian statistical analysis," Charlie was silent about his method. But, almost certainly, this is what he must have done.

The problem, remember, is that there is a large number of reports of sightings, many of them contradictory. Most will be a result of people seeing someone they think looks like the person they saw in the newspaper or on TV. It is not that the informants lack credibility; they are simply mistaken. Therefore the challenge is how to distinguish the correct sightings from the false alarms, especially when you consider that the false alarms almost certainly heavily outnumber the accurate sightings.

The key factor that Charlie can make use of depends on the fact that each report has a time associated with it, the time of the supposed sighting. The accurate reports, all being reports of sightings of the real killer, will refer to locations in the city that follow a geometric pattern, reflecting the movements of one individual. On the other hand, the false reports are likely to refer to locations that are spread around in a fairly random fashion, and are inconsistent with being produced by a single person traveling around. But how can you pick out the sightings that correspond to that hidden pattern?

In a precise way, you cannot. But Bayes' theorem provides a way to assign probabilities to the various sightings so that the higher the probability, the more likely that particular sighting is to be correct. Here is how Charlie will have done it.

Picture a map of Los Angeles. The goal is to assign to each grid square on the map whose coordinates are i, j, a probability figure p(i,j,n) that assesses the probability that the killer is in grid square (i,j) at time n. The idea is to use Bayes' theorem to repeatedly update the probabilities p(i,j,n) over time (that is, as n increases), say in five-minute increments.

To start the process off, Charlie needs to assign initial prior probabilities to each of the grid squares. Most likely he determines these probabilities based on the evidence from the recaptured prisoner as to where and when the two separated. Without such information, he could simply assume that the probabilities of the grid squares are all the same.

At each subsequent time point, Charlie calculates the new posterior probability distribution as follows. He takes each new report—a sighting in grid square (i,j) at time n+1—and on the basis of that sighting updates the probability of every grid square (x,y), using the likelihood of that sighting if the killer was in grid square (x,y) at time n. Clearly, for (x,y) = (i,j), Charlie calculates a high likelihood for the sighting at time n+1, particularly if the sighting report says that the killer was doing something that would take time, such as eating a meal or having a haircut.

If (x,y) is near to (i,j), the likelihood Charlie calculates for the killer being in square (i,j) at time n+1 is also high, particularly if the sighting reported that the killer was on foot, and hence unlikely to move far within a five-minute time interval. The exact probability Charlie assigns may vary depending on what the sighting report says the individual was doing. For example, if the individual was reported as "driving north on Third Street" at time n, then Charlie gives the grid squares farther north on Third a higher likelihood of sightings at time n+1 than squares elsewhere.

The probabilities Charlie assigns are also likely to take account of veracity estimations. For example, a report from a bank guard, who gives a fairly detailed description, is more likely to be correct than one from a drunk in a bar, and hence Charlie will assign higher probabilities based on the former than on the latter. Thus, the likelihood for the killer being at square (x,y) at time n+1 based on a high-quality report of him being at square (i,j) at time n is much higher if (x,y) is close to (i,j) than if the two were farther apart, whereas for a low-quality report the likelihood of getting a report of a sighting at square (i,j) is more "generic" and less dependent on (x,y).

Most likely Charlie also takes some other factors into account. For example, a large shopping mall on a Sunday afternoon will likely generate more false reports than an industrial area on a Tuesday night.

This process is, of course, heavily based on human judgments and estimates. On its own, it would be unlikely to lead to any useful conclusion. But this is where the power of Bayes' method comes into play. The large number of sightings, which at first seemed like a problem, now becomes a significant asset. Although the probability distribution Charlie assigns to the map at each time point is highly subjective, it *is* based on a reasonable rationale, and the mathematical precision of Bayes' theorem, when applied many times over, eventually overcomes the vagueness inherent in any human estimation. In effect, what the repeated application of Bayes' theorem does is tease out the underlying pattern in the sightings data that comes from the fact that sightings of the killer were all of the same individual as he moved through the city.

In other words, Bayes' paradigm provides Charlie with a sound quantitative way of simultaneously considering all possible locations at every point in time. Of course, what he gets is not a single "X marks the spot" on the map, but a probability distribution. But as he works through the process, he may reach some stage where high probabilities are assigned to two or three reasonably plausible locations based on recent reports of sightings. If he then gets one or two high-quality reports that dovetail well, Bayes' formula could yield a high probability to one of those locations. And at that point he would contact his brother Don and say, "Get an agent over there now!"

CHAPTER

7 DNA Profiling

We read a lot about DNA profiling these days, as a method used to identify people. Although the technique is often described as "DNA fingerprinting," it has nothing to do with fingerprints. Rather, the popular term is analogous to an older, more established means of identifying people. Although both methods are highly accurate, in either case care has to be taken in calculating the likelihood of a false identification resulting from two different individuals having fingerprints (of either variety) that the test cannot distinguish. And that is where mathematics comes into the picture.

UNITED STATES OF AMERICA V. RAYMOND JENKINS

On June 4, 1999, police officers in Washington, D.C., found the body of Dennis Dolinger, age 51, at his home in Capitol Hill. He had been stabbed multiple times—at least twenty-five according to reports—with a screwdriver that penetrated his brain.

Dolinger had been a management analyst at the Washington Metropolitan Area Transit Authority. He had lived in Capitol Hill for twenty years and was active in the community. He had a wide network of friends and colleagues across the city. In particular, he was a neighborhood politician and had taken a strong stand against drug dealing in the area.

Police found a blood trail leading from the basement where Dolinger was discovered to the first and second floors of his house and to the front walkway and sidewalk. Bloody clothing was found in the basement and in a room on the second floor. Police believed that some of the bloodstains were those of the murderer, who was cut during the assault. Dolinger's wallet, containing cash and credit cards, had been taken, and his diamond ring and gold chain were missing.

The police quickly identified several suspects: Dolinger's former boyfriend (Dolinger was openly gay), who had assaulted him in the past and had left the D.C. area around the time police discovered the body; a man who was observed fleeing from Dolinger's house but did not call the police; neighborhood drug dealers, including one in whose murder trial Dolinger was a government witness; neighbors who had committed acts of violence against Dolinger's pets; various homeless individuals who frequently visited Dolinger; and gay men whom Dolinger had met at bars through Internet dating services.

By far the strongest lead was when a man named Stephen Watson used one of Dolinger's credit cards at a hair salon and department store in Alexandria within fifteen hours of Dolinger's death. Watson was a drug addict and had a long criminal record that included drug offenses, property offenses, and assaults. Police spoke with a witness who knew Watson personally and saw him on the day of the murder in the general vicinity of Dolinger's home, "appearing nervous and agitated," with "a cloth wrapped around his hand," and wearing a "T-shirt with blood on it." Another witness also saw Watson in the general vicinity of Dolinger's home on the day of the murder, and noted that Watson had several credit cards with him.

On June 9, police executed a search warrant at Watson's house in Alexandria, Virginia, where they found some personal papers belonging to Dolinger. They also noticed that Watson, who was present during the search, had a cut on his finger "that appeared to be several days old and was beginning to heal." At this point, the police arrested him. When questioned at the police station, Watson "initially denied knowing the decedent and using the credit card" but later claimed that "he found a wallet in a backpack by a bank alongside a beige-colored tarp and buckets on King Street" in Alexandria. Based on those facts, the police charged Watson with felony murder.

That might seem to be the end of the matter—a clear-cut case, you might think. But things were about to become considerably more complicated. The FBI had extracted and analyzed DNA from various blood samples collected from the crime scene and none of it matched that of

Watson. As a result, the U.S. Attorney's Office dropped the case against Watson, who was released from custody.

At this point, we need to take a look at the method of identification using DNA, a process known as DNA profiling.

DNA PROFILING

The DNA molecule comprises two long strands, twisted around each other in the now familiar double-helix structure, joined together in a rope-ladder-fashion by chemical building blocks called bases. (The two strands constitute the "ropes" of the "ladder," the bonds between the bases its "rungs.") There are four different bases, adenine (A), thymine (T), guanine (G), and cytosine (C). The human genome is made of a sequence of roughly three billion of these base-pairs. Proceeding along the DNA molecule, the sequence of letters denoting the order of the bases (a portion might be . . . AATGGGCATTTTGAC . . .) provides a "readout" of the genetic code of the person (or other living entity). It is this "readout" that provides the basis for DNA profiling.

Every person's DNA is unique; if you know the exact, three-billion-long letter sequence of someone's DNA, you know who that person is, with no possibility of error. However, using today's techniques, and most likely tomorrow's as well, it would be totally impractical to do a DNA identification by determining all three billion letters. What is done instead is an examination of a very small handful of sites of variation, and the use of mathematics to determine the accuracy of the resulting identification.

DNA is arranged into large structural bodies called chromosomes. Humans have twenty-three pairs of chromosomes which together make up the human genome. In each pair, one chromosome is inherited from the mother and one from the father. This means that an individual will have two complete sets of genetic material. A "gene" is really a location (locus) on a chromosome. Some genes may have different versions, which are referred to as "alleles." A pair of chromosomes have the same loci along their entire length, but may have different alleles at some of the loci. Alleles are characterized by their slightly different base sequences and are distinguished by their different phenotypic effects. Some of the genes studied in forensic DNA tests have as many as thirty-five different alleles.

Most people share very similar loci, but some loci vary from person to person with high frequency. Comparing variations in these loci allows scientists to answer the question of whether two different DNA samples come from the same person. If the two profiles match at each of the loci examined, the profiles are said to match. If the profiles fail to match at one or more loci, then the profiles do not match, and it is virtually certain that the samples do not come from the same person.*

A match does not mean that the two samples must absolutely have come from the same source; all that can be said is that, so far as the test was able to determine, the two profiles were identical, but it is possible for more than one person to have the same profile across several loci. At any given locus, the percentage of people having matching DNA fragments is small but not zero. DNA tests gain their power from the conjunction of matches at each of several loci; it is extremely rare for two samples taken from unrelated individuals to show such congruence over many loci. This is where mathematics gets into the picture.

THE FBI'S CODIS SYSTEM

In 1994, recognizing the growing importance of forensic DNA analysis, Congress enacted the DNA Identification Act, which authorized the creation of a national convicted offender DNA database and established the DNA Advisory Board (DAB) to advise the FBI on the issue.

CODIS, the FBI's DNA profiling system (the name stands for COmbined DNA Index System) had been started as a pilot program in 1990. The system weds computer and DNA technologies to provide a powerful tool for fighting crime. The CODIS DNA database comprises four categories of DNA records:

- Convicted Offenders: DNA identification records of persons convicted of crimes
- Forensic: analyses of DNA samples recovered from crime scenes

---

*The comparison is not made directly between the sequences of the four base letters, but on numerical counts of them. The "DNA profile" is actually a sequence of those counts. The distinction is not important for our account.

- Unidentified Human Remains: analyses of DNA samples recovered from unidentified human remains
- Relatives of Missing Persons: analyses of DNA samples voluntarily contributed by relatives of missing persons

The CODIS database of convicted offenders currently contains in excess of 3 million records.

The DNA profiles stored in CODIS are based on thirteen specific loci, selected because they exhibit considerable variation among the population.

CODIS utilizes computer software to automatically search these databases for matching DNA profiles. The system also maintains a population file, a database of anonymous DNA profiles used to determine the statistical significance of a match.

CODIS is not a comprehensive criminal database, but rather a system of pointers; the database contains only information necessary for making matches. Profiles stored in CODIS contain a specimen identifier, the sponsoring laboratory's identifier, the initials (or name) of DNA personnel associated with the analysis, and the actual DNA characteristics. CODIS does not store criminal-history information, case-related information, social security numbers, or dates of birth.

When two randomly chosen DNA samples match completely in a large number of regions, such as the thirteen used in the CODIS system, the probability that they could have come from two unrelated people is virtually zero. This fact makes DNA identification extremely reliable (when performed correctly). The degree of reliability is generally measured by using probability theory to determine the likelihood of finding a particular profile among a random selection of the population.

BACK TO THE JENKINS CASE

With their prime suspect cleared because his DNA profile did not match any found at the crime scene, the FBI ran the crime scene DNA profile through the CODIS database to see if a match could be found, but the search came out negative.

Six months later, in November 1999, the DNA profile of the unknown contributor of the blood evidence was sent to the Virginia Division of Forensic Science, where a computer search was carried out to compare the profile against the 101,905 offender profiles in its databank. This time a match was found—albeit at only eight of the thirteen CODIS loci, since the Virginia database, being older, listed profiles based on those eight loci only.

The eight-loci match was with a man listed as Robert P. Garrett. A search of law enforcement records revealed that Robert P. Garrett was an alias used by Raymond Anthony Jenkins, an African-American who was serving time in prison for second-degree burglary—a sentence imposed following his arrest in July 1999, a few weeks after Dolinger was murdered. From that point on, the police investigation focused only on Jenkins.

On November 18, 1999, police interviewed a witness—a man who was in police custody at the time with several cases pending against him—who claimed to know Jenkins. This witness reported that on the day after Dolinger's death he had seen Jenkins with several items of jewelry, including a ring with diamonds and some gold chains, and more than $1,000 in cash. Jenkins also appeared to have numerous scratches or cuts to his face, according to government documents.

Seven days later the police executed a search warrant on Jenkins and obtained blood samples. The samples were sent to the FBI's forensic science lab for comparison. In late December 1999, Jenkins' samples were analyzed and profiled on the FBI's thirteen CODIS loci, the eight used by the Virginia authorities plus five others. According to a police affidavit, the resulting profile was "positively identified as being the same DNA profile as that of the DNA profile of the unknown blood evidence that was recovered from the scene of the homicide." The FBI analysis identified Jenkins' blood on a pair of jeans found in the basement near Dolinger, a shirt found in the upstairs exercise room, a towel on the basement bathroom rack, the sink stopper in the sink of the same bathroom, and a railing between the first and second floors of the residence. The FBI estimated that the probability that a random person selected from the African-American population would share Jenkins' profile is 1 in 26 quintillion. Based on that information, an arrest warrant was issued, and Jenkins was arrested on January 13, 2000.

In April 2000, Raymond Jenkins was formally charged with second-degree murder while armed and in possession of a prohibited weapon, a charge that was superseded in October of the same year by onc of two counts of felony murder and one count each of first-degree premeditated murder, first-degree burglary while armed, attempted robbery while armed, and the possession of a prohibited weapon.

Such is the power of DNA profiling, one of the most powerful weapons in the law enforcement agent's arsenal. Yet, as we shall see, that power rests on mathematics as much as on biochemistry, and that power is not obtained without some cost.

THE MATH OF DNA PROFILING

By way of an introductory example, consider a profile based on just three sites. The probability that someone would match a random DNA sample at any one site is roughly one in ten (1/10).* So the probability that someone would match a random sample at three sites would be about one in a thousand:

$$1/10 \times 1/10 \times 1/10 = 1/1{,}000$$

Applying the same probability calculation to all thirteen sites used in the FBI's CODIS system would mean that the chances of matching a given DNA sample at random in the population are about one in 10 trillion:

$$(1/10)^{13} = 1/10{,}000{,}000{,}000{,}000$$

This figure is known as the random match probability (RMP). It is computed using the product rule for multiplying probabilities, which is valid

*Profile match probabilities are based on empirical studies of allele frequencies of large numbers of samples. The figure 1/10 used here is widely regarded as being a good representative figure.

only if the patterns found in two distinct sites are independent. During the early days of DNA profiling, this was a matter of some considerable debate, but for the most part that issue seems to have largely, though not completely, died away.

In practice, the actual probabilities vary, depending on several factors, but the figures calculated above generally are taken to be a fairly reliable indicator of the likelihood of a random match. That is, the RMP is accepted as a good indicator of the rarity of a particular DNA profile in the population at large, although this interpretation needs to be viewed with care. (For example, identical twins share almost identical DNA profiles.)

The denominator in the FBI's claimed figure of 1 in 26 quintillion in the Jenkins case seems absurdly high, and really of little more than theoretical value, when you consider the likelihood of other errors, such as data entry mistakes, contamination errors during sample collection, or laboratory errors during the analysis process.

Nevertheless, whatever actual numbers you compute, it is surely the case that a DNA profile match on all thirteen of the sites used by the FBI is a virtual certain identification—*provided that the match was arrived at by a process consistent with the randomness that underpins the RMP*. As we shall see, however, the mathematics is very sensitive to how well that assumption is satisfied.

USING DNA PROFILING

Suppose that, as often occurs, the authorities investigating a crime obtain evidence that points to a particular individual as the criminal, but fails to identify the suspect with sufficient certainty to obtain a conviction. If the suspect's DNA profile is in the CODIS database, or if a sample is taken and a profile prepared, it may be compared with a profile taken from a sample collected at the crime scene. If the two profiles agree on all thirteen loci, then for all practical—and all legal—purposes, the suspect can be assumed to have been identified with certainty. The random match probability (1 in 10 trillion) provides a reliable estimate of the likelihood that the two profiles came from different individuals. (The one caveat is that relatives should be eliminated. This is not always easy, even for close relatives such as siblings; brothers and sisters are sometimes separated at birth and may not be aware that they have a sibling, and official records do not always correspond to reality.)

Of course, all that a DNA match does is identify—within a certain degree of confidence—an individual whose DNA profile was the same as that of a sample (or samples) found at the crime scene. It does not imply that the individual committed the crime. Other evidence is required to do that. For example, if semen taken from the vagina of a woman who was raped and murdered provides a DNA profile match with a particular individual, then, within the calculated accuracy of the DNA matching procedure, it may be assumed that the individual had sex with the woman not long before her death. Other evidence would be required to conclude that the man raped the woman, and possibly further evidence still that he subsequently murdered her. A DNA match is only that: a match of two profiles.

As to the degree of confidence that can be vested in the identification of an individual by means of a DNA profile match obtained in the above manner, the issues to be considered are:

- The likelihood of errors in collecting or labeling the two samples and determining the associated DNA profiles
- The likelihood that the profile match is purely coincidental*

A likelihood of 1 in 10 trillion attached to the second of these two possibilities (such as is given by the RMP for a thirteen-loci match) would clearly imply that the former possibility is far more likely, since hardly any human procedure can claim a one-in-ten-trillion fallibility rate. Put differently, if there is no reason to doubt the accuracy of the sample collection procedures and the laboratory analyses, the DNA profile identification could surely be viewed with considerable confidence. Provided, that is, the match is arrived at by comparing a profile from a sample from the crime scene with a profile taken from a sample from a suspect

*As will be explained later, care is required in interpreting this requirement in terms of exactly what numerical probability is to be computed.

who has already been identified by means other than his or her DNA profile. But this is not what happened in the Jenkins case. There, Jenkins became a suspect solely as a result of investigators trawling through a DNA database (two databases, in fact) until a match was found—the so-called "cold hit" process.

And that brings in a whole different mathematical calculation.

COLD HIT SEARCHES

In general, a search through a DNA database, carried out to see if a profile can be found that matches the profile of a given sample—say, one obtained from a crime scene—is called a cold hit search. A match that results from such a search would be considered "cold" because prior to the match the individual concerned was not a suspect.

For example, CODIS enables government crime laboratories at a state and local level to conduct national searches that might reveal that semen deposited during an unsolved rape in Florida could have come from a known offender from Virginia.

As in the case where DNA profiling is used to provide identification of an individual who was already a suspect, the principal question that should be asked after a cold hit search has led to a match is: Does the match indicate that the profile in the database belongs to the same person whose sample formed the basis of the search, or is the match purely coincidental? At this point, the mathematical waters rapidly become unexpectedly murky.

To illustrate the problems inherent in the cold hit procedure, consider the following analogy. In a typical state lottery, the probability of winning a major jackpot is around 1 in 35,000,000. To any single individual, buying a ticket is clearly a waste of time. Those odds are effectively nil. But suppose that each week, at least 35,000,000 people actually do buy a ticket. (This is a realistic example.) Then, every one to three weeks, on average, someone will win. The news reporters will go out and interview that lucky person. What is special about that person? Absolutely nothing. The only thing you can say about that individual is that he or she is the one who had the winning numbers. You can make absolutely no other conclusion. The 1 in 35,000,000 odds tell you nothing about any other feature of that person. The fact that there is a winner reflects the fact that 35,000,000 people bought a ticket—and nothing else.

Compare this to a reporter who hears about a person with a reputation of being unusually lucky, accompanies them as they buy their ticket, and sits alongside them as they watch the lottery result announced on TV. Lo and behold, that person wins. What would you conclude? Most likely, that there has been a swindle. With odds of 1 in 35,000,000, it's impossible to conclude anything else in this situation.

In the first case, the long odds tell you nothing about the winning person, other than that they won. In the second case, the long odds tell you a lot.

A cold hit measured by RMP is like the first case. All it tells you is that there is a DNA profile match. It does not, in and of itself, tell you anything else, and certainly not that that person is guilty of the crime.

On the other hand, if an individual is identified as a crime suspect by means other than a DNA match, then a subsequent DNA match is like the second case. It tells you a lot. Indeed, assuming the initial identification had a rational, relevant basis (such as a reputation for being lucky in the lottery case), the long RMP odds against a match could be taken as conclusive. But as with the lottery example, in order for the long odds to have any weight, the initial identification has to be *before* the DNA comparison is run (or at least demonstrably independent thereof). Do the DNA comparison first, and those impressive-sounding long odds could be meaningless.

NRC I AND NRC II

In 1989, eager to make use of the newly emerging technology of DNA profiling for the identification of suspects in a criminal case, including cold hit identifications, the FBI urged the National Research Council to carry out a study of the issue. The NRC formed the Committee on DNA Technology in Forensic Science, which issued its report in 1992. Titled *DNA Technology in Forensic Science*, and published by the National Academy Press, the report is often referred to as NRC I. The committee's main recommendation regarding the cold hit process was:

> The distinction between finding a match between an evidence sample and a suspect sample and finding a match between an evidence sample and one of many entries in a DNA profile databank is important. The chance of finding a match in the second case is considerably higher. . . . The initial match should be used as probable cause to obtain a blood sample from the suspect, but only the statistical frequency associated with the additional loci should be presented at trial (to prevent the selection bias that is inherent in searching a databank).

In part because of the controversy the NRC I report generated among scientists regarding the methodology proposed, and in part because courts were observed to misinterpret or misapply some of the statements in the report, in 1993 the NRC carried out a follow-up study. A second committee was assembled, and it issued its report in 1996. Often referred to as NRC II, the second report, *The Evaluation of Forensic DNA Evidence*, was published by National Academy Press in 1996. The NRC II committee's main recommendation regarding cold hit probabilities was:

> When the suspect is found by a search of DNA databases, the random-match probability should be multiplied by N, the number of persons in the database.

The statistic that NRC II recommends using is generally referred to as the "database match probability," or DMP. This is an unfortunate choice of name, since the DMP is *not* a probability—although in all actual instances it is a number between 0 and 1, and it does (in the view of the NRC II committee) provide a good indication of the likelihood of getting an accidental match when a cold hit search is carried out. (The intuition is fairly clear. In a search for a match in a database of N entries, there are N chances of finding such a match.) For a true probability measure, if an event has probability 1, then it is certain to happen. However, consider a hypothetical case where a DNA database of 1,000,000 entries is searched for a profile having an RMP of 1/1,000,000. In that case, the DMP is

$$1{,}000{,}000 \times 1/1{,}000{,}000 = 1$$

However, in this case the probability that the search will result in a match is not 1 but approximately 0.6312.

The committee's explanation for recommending the use of the DMP to provide a scientific measure of the accuracy of a cold hit match reads as follows:

> A special circumstance arises when the suspect is identified not by an eyewitness or by circumstantial evidence but rather by a search through a large DNA database. If the only reason that the person becomes a suspect is that his DNA profile turned up in a database, the calculations must be modified. There are several approaches, of which we discuss two. The first, advocated by the 1992 NRC report, is to base probability calculations solely on loci not used in the search. That is a sound procedure, but it wastes information, and if too many loci are used for identification of the suspect, not enough might be left for an adequate subsequent analysis. . . . A second procedure is to apply a simple correction: Multiply the match probability by the size of the database searched. This is the procedure we recommend.

This is essentially the same logic as in our analogy with the state lottery. In the Jenkins case, the DMP associated with the original cold hit search of the eight-loci Virginian database (containing 101,905 profiles) would be (approximately)

$$100{,}000 \times 1/100{,}000{,}000 = 1/1{,}000$$

With such a figure, the likelihood of an accidental match in a cold hit search is quite high (recall the state lottery analogy). Thus, what seemed at first like a clear-cut case suddenly begins to look less so. That's what the courts think, too. At the time of writing, the Jenkins case is still going through the legal system, having become one of several test cases across the country.

NUMBERS IN COURT: THE STATISTICAL OPTIONS

So far, the courts have shown reluctance for juries to be presented with the statistical arguments involved in cold hit DNA cases. This is reasonable. To date, experts have proposed at least five different procedures to calculate the probability that a cold hit identification produces a false positive, that is, identifies someone who, by pure happenstance, has the same profile as the sample found at the crime scene. The five procedures are:

1. *Report the RMP alone.* While some statisticians have argued in favor of this approach, many have argued strongly against it. The NRC II report came down firmly against any mention of the RMP in court.
2. *Report the DMP alone.* This is the approach advocated by NRC II.
3. *Report both the RMP and the DMP.* This approach is advocated by the FBI's DNA Advisory Board, which argues that both figures are "of particular interest" to the jury in a cold hit case, although it's not clear how laypersons could weigh the relative significance of the two figures. Nor indeed is it at all clear that it would be right to ask them to so do, when some of the world's best statisticians are not agreed on the matter.
4. *Report the results of an alternative Bayesian analysis.* Some statisticians argue that the issue of assigning a probability to a cold hit identification should be tackled from a Bayesian perspective. (See Chapter 6 for a discussion of Bayesian statistics.) Using Bayesian analysis to compute a reliability statistic for a cold hit match leads to a figure just slightly smaller than the RMP.
5. *Report the RMP calculated on confirmatory loci not considered in the initial search.* This is the approach advocated by NRC I.

At this point, most laypeople are likely to say, "Look, since DNA profiling has an inaccuracy rate of less than one in many trillions (or more), the chances of there being a false match in a database of maybe 3 million entries, such as the CODIS database, is so tiny that no matter which method you use to calculate the odds, a match will surely be definitive proof." The intuition behind such a conclusion is presumably that the database search has 3 million shots at finding a match, so if the odds against there being a match are 1 in 10 trillion, then the odds against finding a match in the entire database are roughly 1 in 3 million (3 million divided by 3 trillion is roughly 1/3,000,000).

Unfortunately—at least it could be unfortunate for an innocent defendant in the case—this argument is not valid. In fact, notwithstanding an RMP in the "one in many trillions" range, even a fairly small DNA database is likely to contain numerous pairs of accidental matches, where two different people have the same DNA profile. A tiny RMP simply does not mean there won't be accidental matches. This is a more subtle version of the well-known birthday puzzle that says you need only have 23 randomly selected people in a room for there to be a better-than-even chance that two of them will have the same birthday. (The exact calculation is a bit intricate, but you get a sense of what is going on when you realize that with 23 people, there are $23 \times 22 = 506$ possible pairs of people, each of which might share a birthday, and that turns out to be just enough pairs to tilt the odds to .508 in favor of there being a match.)

For example, the Arizona DNA convicted offender database is a fairly small one, with some 65,000 entries, each being a thirteen-loci profile. Suppose, for simplicity, that the probability of a random match at a single locus is 1/10, a figure that, as we observed earlier, is not unreasonable. Thus, the RMP for a nine-locus match is $1/10^9$, i.e., 1 in 1 billion. You might think that with such long odds against a randomly selected pair of profiles matching at nine loci, it would be highly unlikely that the database contained a pair of entries that were identical on nine loci. Yet, by an argument similar to the one used in the birthday puzzle, the probability of getting two profiles that match on nine loci is around 5 percent, or 1 in 20. For a database of 65,000 entries, that means you would be quite likely to find some matching profiles!

We'll sketch the calculation at the end of the chapter, but the answer becomes less surprising when you realize that for a database of 65,000 entries, there are roughly $65{,}000^2$—that is, 4,225,000,000—possible pairs of entries, each of which has a chance of yielding a nine-loci match.

In 2005, an actual analysis of the Arizona database uncovered 144 individuals whose DNA profiles matched at nine loci. There were another few that matched at ten loci, one pair that matched at eleven, and one pair that matched at twelve. The eleven and twelve loci matches turned out to be siblings, hence not random. But the others were not, and were, in fact, close to what one should expect from the mathematics when you replace our simplifying 1/10 single-locus match assumption with a realistic figure obtained empirically.

All of which leaves judges and juries facing a mathematical nightmare in reasoning their way to a just decision. On the other hand, even after the mathematical complexities are taken into account, DNA profiling is considerably more reliable than that much older identification standby: fingerprints, which we look at in Chapter 9.

The Database Match Calculation

Here is the calculation we promised earlier. Recall that we have a DNA profile database with 65,000 entries, each entry being a thirteen-loci profile. We suppose that the probability of a random match at a single locus is 1/10, so the RMP for a nine-locus match is $1/10^9$, that is 1 in a billion.

Now, there are $13!/[9! \times 4!] = [13 \times 12 \times 11 \times 10]/[4 \times 3 \times 2 \times 1] = 715$ possible ways to choose nine loci from thirteen, so the RMP for finding a match on *any* nine loci of the thirteen is $715/10^9$.

If you pick any profile in the database, the probability of a second profile not matching on nine loci is roughly $1 - 715/10^9$.

Hence, the probability of all 65,000 database entries not matching on nine loci is roughly $(1 - 715/10^9)^{65,000}$. Using the binomial theorem, this is approximately $1 - 65,000 \times 715/10^9 = 1 - 46,475/10^6$, roughly $1 - .05$.

The probability of there being a nine-loci match is the difference between 1 and this figure, namely $1 - (1 - 0.05) = 0.05$.

CHAPTER

8 Secrets—Making and Breaking Codes

PRIME SUSPECT

In the fifth episode of the first season of *NUMB3RS*, titled "Prime Suspect," broadcast February 18, 2005, a five-year-old girl is kidnapped. Don asks for Charlie's help when he discovers that the girl's father, Ethan, is also a mathematician. When Charlie sees the mathematics Ethan has scribbled on the whiteboard in his home office, he recognizes that Ethan is working on Riemann's Hypothesis, a famous math problem that has resisted attempts at solution for more than 150 years.

The Riemann problem is one of the so-called Millennium Problems, a list of seven unsolved mathematics problems drawn up by an international panel of experts in the year 2000, for each of which the solver will be awarded a $1 million prize. In the case of the Riemann problem, a solution is likely to lead to more than a $1 million prize. It could also lead to a major breakthrough in how to factor large numbers into primes, and hence provide a method for breaking the security code system used to encrypt Internet communications. If that were to happen, Internet commerce would break down immediately, with major economic consequences.

When Don is able to determine the identity of one of the kidnappers, and learns that the plan is to "unlock the world's biggest financial secret" it becomes clear why Ethan's daughter was kidnapped. The captors want to use Ethan's method to break into a bank's computer and steal millions of dollars. Don's obvious strategy is for Ethan to provide the gang with the key to get into the bank's computer and trace the activity electronically in order to catch the thieves. But when Charlie finds a major error in Ethan's argument, the only hope Don has to rescue Ethan's daughter is to come up with a way to fool the kidnappers into believing that he really can provide the Internet encryption key they are demanding, and use that to trace their location to rescue the daughter.

At one point in the episode, Charlie gives a lecture to the FBI agents on how Internet encryption depends on the difficulty of factoring large numbers into primes. Elsewhere in the story, Charlie and Ethan discuss the feasibility of turning Ethan's solution into an algorithm and Charlie refers to "the expansion of the zero-free region to the critical strip." Charlie also observes that the kidnappers would need a supercomputer to factor a large number into primes. Amita, his student, notes that it is possible to build a supercomputer with a large number of PCs linked together. As always, these are all mathematically meaningful and realistic statements. So too is the basic premise for the story: a solution to the Riemann problem might very well lead to a collapse of methods currently used to keep Internet communications secure. Ever since the Second World War, message encryption has been the business of mathematicians.

WWW.CYBERCRIME.GOV

These days, you don't need a gun or a knife to steal money. A cheap personal computer and an Internet connection will do. It's called cybercrime; it's a new form of crime; it is substantial; and it is growing. It includes a broad range of illegal activities, such as software piracy, music piracy, credit card fraud (of many kinds), identity theft, manipulation of stocks, corporate espionage, child pornography, and "phishing" (sending a computer user an e-mail that purports to be from a financial institution that seeks to trick the receiver into revealing their bank details and other personal data).

There are no reliable figures on the extent of cybercrime, since many banks and Internet commerce companies keep such information secret, to avoid giving the impression that your money or credit card number is not safe in their hands. It has been suggested, though hotly disputed, that the annual proceeds from cybercrime may be in excess of $100 billion. If that were true, it would exceed the sale of illegal drugs. Regardless of the actual figures, cybercrime is a sufficiently major problem that both the U.S. Department of Justice and the FBI have entire units that focus on such criminal activity, and both have websites devoted to information about it: www.cybercrime.gov and www.fbi.gov/cyberinvest/cyberhome.htm, respectively.

The 2005 FBI computer crime survey, developed and analyzed with the help of leading public and private authorities on cyber security, and based on responses from a cross section of more than 2,000 public and private organizations in four states, reported that:

- Nearly nine out of ten organizations experienced computer security incidents in the year; 20 percent of them indicated they had experienced twenty or more attacks; viruses (83.7 percent) and spyware (79.5 percent) headed the list.
- Over 64 percent of the respondents incurred a financial loss. Viruses and worms cost the most, accounting for $12 million of the $32 million in total losses.
- The attacks came from thirty-six different countries. The United States (26.1 percent) and China (23.9 percent) were the source of more than half of the intrusion attempts, though many attackers route through one or more intermediate computers in different countries, which makes it difficult to get an accurate reading.

Law enforcement agents who focus their energies on cybercrime use mathematics in much of their work. In many cases, they use the same techniques as are described elsewhere in this book. In this chapter, however, we'll focus our attention on one important aspect of the fight against cybercrime that uses different mathematics, namely Internet security. In this area, ingenious use of some sophisticated mathematics has led to major advances, with the result that, if properly used, the systems available today for keeping Internet communications secure are extremely reliable.

KEEPING SECRETS

When you use an ATM to withdraw money from your account, or send your credit card details to an Internet retailer, you want to be sure that only the intended receiver has access to the details you send. This cannot be achieved by preventing unauthorized third parties from "eavesdropping" on the electronic messages that pass between you and the organization you are dealing with. The Internet is what is called an open system, which means that the connections between the millions of computers that make up the network are, to all intents and purposes, public. Security of Internet communications traffic is achieved by means of encryption—"scrambling" the message so that, even if an unauthorized third party picks up the signal transmitted, the eavesdropper will be unable to make sense of it.

The notion of encryption is not new. The idea of using a secret code to keep the contents of a message secret goes back at least as far as the days of the Roman Empire, when Julius Caesar used secret codes to ensure the security of the orders he sent to his generals during the Gallic wars. In what is nowadays called a Caesar cipher, the original message is transformed by taking each letter of each word in turn and replacing it by another letter according to some fixed rule, such as taking the letter three places along in the alphabet, so A is replaced by D, G by J, Y by B, and so on. Thus the word "mathematics" would become "pdwkhpdwlfv".

A message encrypted using a Caesar cipher may look on the surface to be totally indecipherable without knowing the rule used, but this is by no means the case. For one thing, there are only twenty-five such "shift along" ciphers, and an enemy who suspected you were using one need only try them all in turn until the one used was found.

A slightly more robust approach would be to employ some other, less obvious rule for substituting letters. Unfortunately, any such substitution cipher, which simply replaces one letter by another, is highly vulnerable to being broken by a simple pattern analysis. For instance, there are very definite frequencies with which individual letters occur in English (or in any other language), and by counting the number of occurrences of each letter in your coded text, an enemy can easily deduce just what your substitution rule is—especially when computers are used to speed up the process.

With simple substitution out of the question, what else might you try? Whatever you choose, similar dangers are present. If there is any kind of recognizable pattern to your coded text, a sophisticated statistical analysis can usually crack the code without much difficulty.

To be secure, therefore, an encryption system must destroy any pattern that the enemy could discover in order to break the code. Yet, the transformation performed on the message by your encryption scheme clearly cannot destroy *all* order—the message itself must still be there beneath it all, to allow the intended receiver to recover it. The trick, then, is to design the encryption system so that this hidden order is buried sufficiently deeply to prevent an enemy from discovering it.

All cipher systems employed since the end of the Second World War depend on mathematics, and all use computers. They have to. Because the enemy may be assumed to have powerful computers to analyze your encrypted message, your system needs to be sufficiently complex to resist computer attack.

It takes a lot of time and effort to design and build a secure encryption system. To avoid having constantly to develop new systems, modern encryption systems invariably consist of two components: an encryption procedure and a "key." The former is, typically, a computer program or possibly a specially designed computer. In order to encrypt a message the system requires not only the message but also the chosen key, usually a secret number. The encryption program will code the message in a way that depends upon the chosen key, so that only by knowing that key will it be possible to decode the ciphered text. Because the security depends on the key, the same encryption program may be used by many people for a long period of time, and this means that a great deal of time and effort can be put into its design.

An obvious analogy is that manufacturers of safes and locks are able to stay in business by designing one type of lock which may be sold to hundreds of users, who rely upon the uniqueness of their own key to provide security. (The "key" in this case could be a physical key or a numerical combination.) Just as an enemy may know how your lock is designed and yet still be unable to break into your safe without having the physical key or knowing the combination, so the enemy may know what encryption system you are using without being able to crack your coded messages—a task for which knowledge of your key is required.

In some key-based encryption systems, the message sender and receiver agree beforehand on some secret key that they then use to send each other messages. As long as they keep this key secret the system, if it is well designed, should be secure. One such system used for many years, though now regarded as a bit too long in the tooth and vulnerable to attack using computers much faster than were available when it was first developed, is the American-designed Data Encryption Standard (DES). The DES requires for its key a number whose binary representation has 56 bits (in other words, a string of 56 zeros and ones operates as the key). Why such a long key? Well, no one made any secret of how the DES system works. All the details were published at the outset. That means that an enemy could crack your coded messages simply by trying all possible keys one after the other until one is found which works. With the DES, there are 2^{56} possible keys to be tried, a number that was large enough to render the task virtually impossible in the days when the system was first used.

Encryption systems such as DES have an obvious drawback. Before such a scheme can be used, the sender and receiver have to agree on the key they will use. Since they will not want to transmit that key over any communication channel, they have to meet and choose the key, or at the very least employ a trusted courier to convey the key from one to the other. This is fine for setting up Internet access to your bank account; you can simply go along in person to your local branch and set up the key in person. But it is no use at all to establish secure communication between individuals who have not already met. In particular, it is not suitable for use in Internet commerce, where people want to send secure messages across the world to someone they have never met.

PUBLIC KEY CRYPTOGRAPHY

In 1976, two young researchers at Stanford University, Whitfield Diffie and Martin Hellman, published a landmark paper titled "New Directions in Cryptography," in which they proposed a new type of cipher system: public key cryptography. In a public key system, the encryption method requires not one but two keys—one for enciphering and the other for deciphering. (This would be like having a lock that requires one key to lock it and another to unlock it.) Such a system would be used as follows, they suggested.

An individual, let's call her Alice, who wishes to use the system, purchases the standard program (or special computer) used by all members of the communication network concerned. Alice then generates two keys. One of these, her deciphering key, she keeps secret. The other key, the one that will be used by anyone else on the network for encoding messages they want to send *to her*, she publishes in a directory of the network users.

If another network user, Bob, wants to send Alice a message, he looks up Alice's public enciphering key, encrypts the message using that key, and sends the encrypted message to Alice. To decode the message, it is of no help knowing (as anyone can) Alice's enciphering key. You need the deciphering key. And only Alice, the intended receiver, knows that. (An intriguing feature of such a system is that once Bob has enciphered his message, he cannot decipher it; so if he wants to refer to it later he'd better keep a copy of the original, unciphered version!)

Diffie and Hellman were not able to come up with a reliable way to construct such a system, but the idea was brilliant, and it was not long before three researchers at MIT, Ronal Rivest, Adi Shamir, and Leonard Adleman, found how to make the suggestion work. Their idea was to exploit the strengths and weaknesses of those very computers whose existence makes the encryption-scheme designer's task so difficult.

It turns out that it is relatively easy to write a computer program to find large prime numbers, say, on the order of 150 digits. It is also easy to multiply two such large primes together to produce a single (composite) number of around 300 digits or more. But factoring a number of that size into its component primes is not at all easy, and indeed, to all intents and purposes, is impossible. (More precisely, it would take the fastest computer many decades, or even centuries, to find the factors.) The public key system based on this idea is called the RSA system, after the initials of the three inventors. The success of the method led to the establishment of a commercial company specializing in data security, RSA Data Security, Inc., based in Redwood City, California.

The secret deciphering key used in the RSA method consists essentially of two large prime numbers chosen by the user. (Chosen with the aid of a computer—not taken from any published list of primes, which an enemy might have access to!) The public enciphering key is the product of these two primes. Since there is no known fast method of factoring large numbers, it is practically impossible to recover the deciphering key from the public enciphering key. Message encryption corresponds to multiplication of two large primes (an easy computational task), decryption to the opposite process of factoring (a hard computational task).

We should point out that the encryption is not actually achieved by multiplying primes, nor is decryption carried out by factoring. Rather, that is how the keys are generated. That term "corresponds to" in the above description should be read very loosely. While encryption and decryption are not merely multiplication and factoring, the RSA system is, however, arithmetical. The message is first translated into numerical form, and the encryption and decryption processes consist of fairly simple arithmetical operations performed on numbers.

Clearly, then, the security of the RSA system, and accordingly of the many international data networks that use it, relies upon the *inability* of mathematicians to find an efficient method of factoring large numbers.

As you might expect, with so much at stake, the widespread use of the RSA system has spurred a considerable amount of research into the problems of finding primes and of factoring large numbers.

The obvious way to determine whether a number N is prime is to see if any smaller number divides it. A few moments' thought shows that you need only check to see if any number below or equal to $\sqrt{N}$ divides N. If N is fairly small, say three or four digits, this is feasible by hand; with a standard desktop PC, you could handle numbers with more digits. But the task becomes impractical when N has, say, fifty digits or more. However, there are other ways to check if a number N is prime, which do not require a blind search through all possible factors up to $\sqrt{N}$, and some of them are efficient enough that they can work well on a reasonably fast computer for numbers with hundreds of digits. Thus, finding primes to generate the keys in public key cryptography is not a problem.

The methods actually used to test primality are all beyond the scope of this book, but a simple example will show how you can determine that a number is prime without having to look at and eliminate all possible factors. The example comes from the work of the great French mathematician Pierre de Fermat (1601–65).

Though only an "amateur" mathematician (he was a jurist by profession), Fermat produced some of the cleverest results mathematics has ever seen to this day. One of his observations was that if p is a prime number, then for any number a less than p, the number $a^{p-1} - 1$ is divisible by p. For instance, suppose we take $p = 7$ and $a = 2$. Then

$$a^{p-1} - 1 = 2^{7-1} - 1 = 2^6 - 1 = 64 - 1 = 63$$

and indeed 63 is divisible by 7. Try it yourself for any values of p (prime) and a (less than p). The result is always the same.

So here is a possible way of testing if a number n is prime or not. Compute the number $2^{n-1} - 1$. See if n divides it. If it does not, then n cannot be prime. (Because if n was prime, then by Fermat's observation you would have divisibility of $2^{n-1} - 1$ by n.) But what can you conclude if you find that n does divide $2^{n-1} - 1$? Not, unfortunately, that n has to be prime. (Though this is quite likely to be the case.) The trouble is, while Fermat's result tells us that n divides $2^{n-1} - 1$ whenever n is prime, it does not say that there are no composite numbers with the same property. (It is like saying that all motor cars have wheels; this does not prevent other things having wheels—bicycles, for instance.) And in fact there are nonprimes which do have the Fermat property. The smallest one is 341, which is not prime, as it is the product of 11 and 31. If you were to check (on a computer) you would find that 341 does divide $2^{340} - 1$. (We shall see in a moment that there is no need to calculate 2^{340} in making this check.) Composite numbers that behave like primes as far as the Fermat property is concerned are called pseudoprimes. So if, when you test for primality using the Fermat result, you discover that n does divide $2^{n-1} - 1$, then all you can conclude is that either n is prime or else it is pseudoprime. (In this case the odds are heavily in favor of n actually being prime. For though there are in fact an infinity of pseudoprimes, they occur much less frequently than the real primes. For instance there are only two such numbers under 1,000, and only 245 below one million.)

In using the above test, it is not necessary to calculate the number 2^{n-1}, a number which will be very large for even quite modest values of n. You only need to find out whether or not n divides $2^{n-1} - 1$. This means that multiples of n may be ignored *at any stage of the calculation*. To put it another way, what has to be calculated is the remainder that *would* be left if $2^{n-1} - 1$ was divided by n. The aim is to see whether or not this remainder is zero, but since multiples of n will obviously not affect the remainder, they may be ignored. Mathematicians (and computer programmers) have a standard way of denoting remainders: the remainder left when A is divided by B is written as

A mod B

Thus, for example, 5 mod 2 is 1, 7 mod 4 is 3, and 8 mod 4 is 0.

As an example of the Fermat test, let us apply it to test the number 61 for primality. We need to calculate the number $[2^{60} - 1] \bmod 61$, which can be written equivalently as $[2^{60} \bmod 61] - 1$. If this is not zero, then 61 is not a prime. If it is zero, then 61 is either a prime or a pseudoprime (and in fact is a genuine prime, as we know already). We shall try to avoid calculating the large number 2^{60}. We start with the observation that $2^6 = 64$, and hence $2^6 \bmod 61 = 3$. Then, since $2^{30} = (2^6)^5$, we get

$$2^{30} \bmod 61 = (2^6)^5 \bmod 61 = (3)^5 \bmod 61 = 243 \bmod 61 = 60$$

So,

$$2^{60} \bmod 61 = (2^{30})^2 \bmod 61 = 60^2 \bmod 61 = 3{,}600 \bmod 61 = 1$$

Thus,

$$2^{60} \bmod 61 - 1 = 0$$

Since the final answer here is 0, the conclusion is that 61 is either prime or pseudoprime, as we anticipated.

One of the methods professionals use to find large primes starts with the Fermat test just described and modifies the approach so it cannot be "fooled" by a pseudoprime. The reason we can't describe the method in this book is that it takes considerable effort, and some sophisticated mathematics, to circumvent the pseudoprime problem.

To date, there is no method to factor a large number that is even remotely as efficient as one of the primality testing methods, despite a considerable investment of talent and effort. Research into the problem has not been without some successes, however, and on several occasions mathematicians have come up with ingenious ways to find factors in usefully short computational time. When the RSA system was first put into use, factoring a number of around 120 digits was at the limit of what could be achieved. Improvements both in algorithm design and computer technology have since brought 120-digit numbers into the vulnerable range, so cryptographers have increased the size of RSA keys to well beyond that level. At the moment, many mathematicians believe it probably is not possible to find a method that can factor (in realistic time) numbers of 300 digits or more, so that is regarded as a safe key size.

That developments in factoring do indeed pose a genuine, if potential, threat to RSA codes was illustrated in dramatic fashion in April 1994, when a sophisticated method was used to crack a challenge problem in RSA cryptography that had been posed in 1977. The origin of the problem is itself of interest. In 1977, when Rivest, Shamir, and Adleman proposed their public-key encryption system, it was described by mathematics writer Martin Gardner in the August issue of *Scientific American*, in his popular mathematics column. There, Gardner presented a short message that had been encoded using the RSA scheme, using a 129-digit key resulting from the multiplication of two large primes. The message and the key were produced by researchers at MIT, who offered, through Gardner, $100 to the first person who managed to crack the code. The composite number that was the key to the code became known as RSA–129. At the time, it was thought it would take more than 20,000 years to factor a 129-digit number of its kind, so the MIT group thought their money was safe. Two developments that followed were to result in the solution to the MIT challenge a mere seventeen years later.

The first was the development of so-called quadratic sieve methods for factoring large numbers. A crucial feature of these methods that was to prove significant in factoring RSA–129 was that they effectively broke up the problem into a large number of smaller factorizations—a process that, while still challenging, was at least feasible with a fairly fast computer. The second pivotal development was the Internet. In 1993, Paul Leyland of Oxford University, Michael Graff at Iowa State University, and Derek Atkins at MIT put out a call on the Internet for individuals to volunteer their—and their personal computers'—time for a massive, worldwide assault on RSA–129. The idea was to distribute the various parts of the factorization problem yielded by the quadratic sieve method, and then sit back and wait until enough of those partial results had been found to produce a factorization of RSA–129. (The quadratic sieve method they used did not require all of the smaller subfactorizations to be solved; just enough of them.) Some 600 volunteers, spread around the world, rose to the challenge. Over the next eight months, results came in at the rate of around 30,000 a day. By April 1994, with greater than 8 million individual results to work on, a powerful supercomputer was set the formidable task of looking for a combination of the small factorizations that would yield a factor of RSA–129. It was a mammoth computation, but in the end it was successful. RSA–129 was factored into two primes, one having 64 digits, the other 65. And with it, the original MIT message was decrypted. It read: *The magic words are squeamish ossifrage*. (This is a typical MIT inside joke. The ossifrage is a rare vulture having a wingspan of up to ten feet, whose name means "bone breaker.")

DIGITAL SIGNATURES

Another security issue Whitfield and Hellman addressed in their 1976 paper was: How can a receiver of an electronic document be sure that it actually came from the source it claimed to be from? In the case of written documents, we generally rely on a signature. Public key cryptosystems provide a means for creating an electronic analog of a signature—a digital signature, as it were. The idea is straightforward: You use the public key encryption system in reverse. If Alice wants to send Bob an electronically signed document, she encrypts it using her secret decryption key. When Bob receives the document, he uses Alice's public encryption key to decrypt the message. This will result in gibberish unless the message was encrypted using Alice's decryption key. Since only Alice knows that key, if the result is a readable document, Bob can be sure that it came from Alice.

In fact, a digital signature is a more secure form of authentication than a regular signature. Someone could always copy (either by hand or electronically) your signature from one document to another, but a digital signature is tied to the document itself. The idea of digital signatures is also used to provide digital certificates, verifications provided by a particular website that it is indeed the site it purports to be.

WHAT KEEPS YOUR PASSWORD SAFE?

Even with message encryption, activities such as online banking still have vulnerabilities. One obvious potential weak point is your password. By transmitting your password in encrypted form, an eavesdropper could not obtain it; but if an enemy were able to hack into the computer on which your bank stores its customers' passwords (which it has to do in order to check your attempted login), he or she would immediately have free access to your account. To prevent this happening, your bank does not store your password; rather it stores what is called a *hashed* version.

Hashing is a particular kind of process that takes an input string (such as your password) and generates a new string of a particular size. (It's not strictly speaking an encryption process since it may be impossible to undo the hash.) When you try to log on to your bank account, the bank's computer compares the hashed version of the password you type in with the entry stored in its hashed-passwords file. To make this system work, the hashing function, H, has to have two fairly obvious properties:

1. For any input string x, it should be easy to compute H(x).
2. Given any hash value y, it should be computationally infeasible to find an x such that H(x) = y.

("Computationally infeasible" means it would take the fastest computers more than, say, a human lifetime to carry out the procedure to completion.)

By requirement 2, even if a hacker gained access to the stored login information, he or she would not be able to obtain your password (though without additional controls they would of course be able to access your account on that machine, since it's the hashed version that the receiving server uses for authorization.)

In practice, the people who design hash functions usually demand an additional uniformity feature that facilitates efficient storage of the hashed values of identification information and makes for a faster and easier database-lookup procedure to determine identity:

3. All values produced by H have the same bit-length.

Because of this third condition, in theory there will be many different input strings that produce the same output; in the parlance of the hashing community, there will be "collisions," distinct input strings x and y such that H(x) = H(y). Because access to secure sites is determined (at the site) by examining the incoming hashed login data, one possible weakness of the system is that illegal access to an account does not require that the intruder obtain the account holder's login identity and password; it is sufficient to find *some* input data that generates the same hashed value—that is, to find an input that collides with the legitimate data. In designing an algorithm for a hash function, it is therefore clearly important to make sure that this is extremely unlikely to occur. That gives a fourth requirement:

4. It is a practical impossibility (it is "computationally infeasible") to find a string y that collides with a given string x, that is, for which H(x) = H(y).

Typically, hash functions work by combining (in some systematic way) the bits of the input string (e.g., your login details) with other bits chosen at random, and performing some complex, iterative distillation process that reduces the resulting string down to one of a fixed length (predetermined for the system).

There are dozens of different hash functions in use. The two most widely used are MD5 ("Message Digest algorithm 5"), developed by Ronald Rivest (he of RSA) at MIT in 1991 as one of a series of hash algorithms he designed, and SHA–1 ("Secure Hash Algorithm 1") developed by the National Security Agency in 1995. MD5 produces a hash value of 128 bits, and it would take on average 2^{64} guesses to find a collision. SHA–1 generates a hash string of length 160 bits, and it would require an average of 2^{80} guesses to find a collision. In theory, both methods would seem to offer a high degree of security—provided that the only feasible way to find a collision is by trial and error.

Unfortunately for the digital security world, trial and error is not the only way to make a dent in a hashing system such as SHA–1. During the late 1990s and early 2000s, Xiaoyun Wang, a mathematician at Tsinghua University in Beijing, showed that with ingenuity and a lot of hard work, it was possible to find collisions for some widely used hashing functions. At the Crypto '04 conference in Santa Barbara in 2004, Wang astonished the attendants with her announcement that she had found a way to find a collision for MD5 in just 2^{37} inputs, a staggering reduction in problem size that made MD5 highly vulnerable.

Wang's approach was to input to the algorithm strings that differ by just a few bits and look closely at what happens to them, step by step, as the algorithm operates on them. This led her to develop a "feel" for the kinds of strings that will result in a collision, allowing her to gradually narrow down the possibilities, resulting eventually in her developing a procedure to generate a collision.

Following the announcement at Crypto '04, Wang, together with her colleagues Hongbo Yu and Yiqun Lisa Yin, started work on the crown jewel of current hash functions, SHA–1. This proved a much harder nut to crack, but to the general dismay (and admiration) of the computer security community, at the annual RSA security conference in San Francisco in February 2005, they were able to announce that they had developed an algorithm that could generate two SHA–1 colliding files in just 2^{69} steps.

Wang and her colleagues have not yet cracked SHA–1; they have just produced a method that *could* crack it in far fewer steps than was previously believed possible. That number 2^{69} is still sufficiently high to provide some degree of confidence in the system's security—for now. So too is the even lower number of 2^{63} steps that Wang and other collaborators managed to achieve in the months following the February 2005 announcement. But many in the cryptographic community now believe that the writing is on the wall, and that, as a result of Wang's work, advances in computing speed and power will rapidly render useless all the hashing algorithms currently in use. It won't happen today—experts assure us that our ATM transactions are secure for now. But soon. Commenting on the development to *New Scientist* magazine, Burt Kaliski, the head of RSA Laboratories in Bedford, Massachusetts, declared, "This is a crisis for the research community." Mark Zimmerman, a cryptographer with ICSA Labs in Mechanicsburg, Pennsylvania, put it rather more colorfully: "It's not Armageddon, but it's a good kick in the pants."

CHAPTER

9 How Reliable Is the Evidence?

Doubts about Fingerprints

THE WRONG GUY?

When Don arrives on the scene he finds that the murderer had garroted his victim. It's not a common method, but it reminds Don of a murder committed a year earlier. On that occasion, the FBI's investigation was very successful. After both eyewitness testimony from a police lineup and a fingerprint match identified a man named Carl Howard as the murderer, Howard confessed to the crime, accepted a plea bargain, and went to prison. But the similarities of that earlier murder to the new one are so striking that Don begins to wonder whether they got the wrong guy when they sent Howard to prison. As Charlie helps Don with the investigation of suspects in the new murder, they speculate about the possibility that Howard was an innocent man sent to prison for a crime he did not commit.

This is the story that viewers watched unfold in the first-season episode of *NUMB3RS* called "Identity Crisis," broadcast on April 1, 2005.

A key piece of evidence that sent Howard to prison was a fingerprint from the murder scene—more accurately, part of a thumbprint. The FBI fingerprint examiner was certain of the correctness of her identification of Howard as the source of the crime-scene partial print, which led first Howard's lawyer and then Howard himself to conclude that accepting a plea bargain was the only sensible thing to do. But once

Howard's possible innocence is being considered, Charlie, the mathematician trained to think logically and to demand scientific proof for scientific claims, engages the fingerprint examiner in a discussion:

CHARLIE: How do you know that everyone has their own unique fingerprint?
EXAMINER: The simple answer is that no two people have ever been found to have the same prints.
CHARLIE: Have you examined everyone's print? Everyone on the planet?

The match the examiner made was based on what is called a "partial," a latent fingerprint consisting of ridge marks from only part of the tip of a single finger. So Charlie continues his questioning, asking how often just *a part of a single finger's print* from one person looks like that of another person. The examiner says she doesn't know, prompting Charlie to press her further.

CHARLIE: There's no data available?
EXAMINER: No. We've never done those population surveys.
CHARLIE: But isn't random-match probability the only way you'll ever be able to know, really know, the likelihood of two prints matching?
AGENT REEVES: That's how DNA matches are made.
CHARLIE: That's what gives DNA those "one in a billion" odds. But prints don't have odds?

As usual, Charlie is right on the ball. These days, fingerprint evidence, once regarded as so infallible that forensic scientists would never consider challenging its certainty, is under increasing attack and critical scrutiny in courts across the United States and many other parts of the world.

THE MYTH OF FINGERPRINTS

The twentieth century's most stunning forensic success is probably the establishment of fingerprint identification as the "gold standard" for scientific evidence in criminal prosecutions. Its acceptance as a virtually unchallengeable "clincher" in the courtroom is shown by the terminology often applied to its only current rival, DNA evidence, which is often referred to as "genetic fingerprinting."

When it first appeared, fingerprinting was not immediately seized upon as the magical key to resolving questions about the identification of criminals. It took decades in the United States and Europe to dislodge its predecessor, the Bertillon system.

Invented by a Parisian police clerk in the late nineteenth century, the Bertillon system relied primarily on an elaborate set of eleven carefully recorded anatomical measurements—the length and width of the head, length of the left middle finger, the distance from the left elbow to the tip of the left middle finger, and so on. That system had proved a great success in foiling the attempts of repeat offenders to avoid harsher sentences by passing themselves off under a succession of aliases.

Like Bertillonage, fingerprinting proved to be a reliable method of "verification." A police department could compare a high-quality set of ten fingerprints obtained from "Alphonse Parker," now in custody, with a file of "full sets" of ten fingerprints from previous offenders and perhaps identify Parker as "Frederick McPhee" from his last incarceration. Even more stunning was the possibility of "lifting" fingerprints from surfaces—a table, a window, a glass—at the scene of a crime and using these "latent prints" to individualize the identification of the perpetrator. That is, by searching through a file of cards containing known exemplars, full-set fingerprints of known individuals, investigators could sometimes obtain a match with crime-scene fingerprints and thereby identify the perpetrator. Or they could bring in a suspect, fingerprint him, and compare those prints with the ones lifted from the crime scene. Even though latent fingerprints are often of low quality—smudged, partial (involving only a portion of the tip of the finger), incomplete (involving only one or two fingers, say)—an experienced and skilled fingerprint examiner could still possibly observe enough commonality with an exemplar print set to make a positive identification with enough certainty to offer testimony in court.

Because the chances of a crime-scene investigation yielding accurate measurements of the perpetrator's head-width and the like are all but zero, the advantage of fingerprinting over Bertillonage for investigative work soon became clear. Even as it was being replaced by fingerprinting, however, the Bertillon system was recognized as having one clear advantage of its own: the indexing system that was developed to go with it. Bertillon relied on numerical values for standardized measurements; accordingly, searches of a large card file to determine a possible match with the measurements of a person in custody could be performed in a straightforward way. Fingerprint matching relied on human judgment in side-by-side comparison of the distinguishing features of two prints or sets of prints, which did not lend itself to the same kind of numerically driven efficiency.

With the advent of computers in the mid-twentieth century, however, it became possible to code sets of fingerprints numerically in such a way that a computer could quickly eliminate the great majority of potential matches and narrow the search to a small subset of a large file, so that human examiners could be used for the final individualization—a possible matching of a suspect print with a single exemplar. Indeed, after September 11, 2001, the United States government accelerated efforts to develop rapid computer-assisted methods to compare quickly fingerprint scans of individuals attempting to enter the country against computer databases of fingerprint features of known or suspected terrorists. These computer-assisted methods, known to fingerprint experts as "semi-lights-out systems," rely heavily upon numerically coded summaries of key features of an individual's fingerprints. Exploiting these features makes it possible to offer a human expert, whose final judgment is considered a necessity, at most a handful of exemplars to check for a match.

For prosecution of criminals, the element of human expertise has proved to be critical. Fingerprint examiners, working for agencies such as the FBI or police departments, have varying levels of training and competence, but their presentations in court invariably rest on two pillars:

- The claim that fingerprints are literally unique: No two people, not even identical twins, have ever been found to have identical fingerprints.

- The certainty of the examiner: With "100 percent confidence" (or words to that effect), he or she is certain that the match between the crime-scene prints and the examplar prints of the defendant is correct; they are fingerprints of the same person.

HOW DOES AN EXPERT "MATCH" FINGERPRINTS?

There is no completely specified protocol for matching fingerprints, but experts generally mark up the pictures of the prints in a way something like this:

Crime scene print

Single finger from exemplar

Every skilled and experienced examiner uses a variety of comparisons between prints to make a match. To their credit, they subscribe to an admirably sound principle, the one dissimilarity doctrine, which says that if any difference between the prints is found that cannot be accounted for or explained—say, by a smudge or speck of dirt—then a potential match must be rejected.

The most common testimony relies, however, on the determination of certain features called minutiae—literally, points on the prints where ridgelines end or split in two. These are sometimes called Galton points, in homage to Sir Francis Galton, the pioneering English statistician, whose 1892 book *Finger Prints* established the basic methods for comparing these points on two prints to make an identification. Unfortunately for the practice of fingerprint forensics, no standard has been established—at least in American practice—for the minimum number of points of commonality needed to determine a reliable match. Many a defense lawyer or judge has been frustrated by the lack of any standardization of the number of points: Is twelve a sufficient number? Is eight enough? In Australia and France, the minimum number is twelve. In

Italy it is sixteen. In the United States, rules of thumb (no pun intended) vary from state to state, even from police department to police department. Essentially, the position of fingerprint experts in court seems to have been "I generally require at least X points," where X is never larger than the number in the present case.

FINGERPRINT EXPERTS VERSUS THE LIKES OF CHARLIE EPPES

In recent years there has been a growing chorus of opposition to the courts' formerly routine acceptance of the automatic certainty of matches claimed by fingerprint expert witnesses. Like Charlie Eppes, a number of mathematicians, statisticians, other scientists, and distinguished lawyers—even some judges—have complained in court and in public about the lack of standards for fingerprint evidence, the performance certification of expert examiners, and, most important of all, the lack of scientifically controlled validation studies of fingerprint matching—that is, the lack of any basis for determining the frequency of errors.

Referring to an acronym for the usual methods of fingerprint identification, ACE-V, a federal judge commented:*

> The court further finds that, while the ACE-V methodology appears to be amenable to testing, such testing has not yet been performed.

To experts in the methods of scientific investigation, it is simply mind-boggling to hear fingerprint evidence justified by the "no two are ever the same" claim. That is, at best, the right answer to the wrong question. Even if the one-trillion-plus possible pairings of full-set "exemplar" prints from the FBI's 150-million-set noncriminal database were thoroughly examined by the best human experts and found to satisfy the "no two ever match" claim, the level of assurance provided by that claim alone would be minimal. The right sort of question is this: How often

---

***United States v. Sullivan*, 246 F. Supp. 2d 700, 704 (E.D. Ky. 2003).

are individual experts wrong when they declare a match between a high-quality exemplar of ten fingers and smudged partial prints of two fingers lifted from a crime scene?

There is a compelling irony in the fact that DNA evidence (discussed in Chapter 7), which in the 1980s and '90s only gradually earned its place in the courtroom as "genetic fingerprinting" through scientific validation studies, is now being cited as the standard for validating the claimed reliability of fingerprint evidence. The careful scientific foundation that was laid then, bringing data and hardcore probability theory and statistical analysis to bear on questions about the likelihood of an erroneous match of DNA, has by now established a "single point of comparison"—but a very powerful one—for fingerprint evidence. Charlie's question, "But prints don't have odds?" isn't heard only on TV.

Just after Christmas in 2005, the Supreme Judicial Court in Massachusetts ruled that prosecutors in the retrial of defendant Terry L. Patterson could not present the proposed testimony of an expert examiner matching Patterson's prints with those found on the car of a Boston police detective who was murdered in 1993. The ruling came after the court solicited amicus curiae ("friend of the court") briefs from a variety of scientific and legal experts regarding the reliability of identifications based on "simultaneous impressions." Specifically, the examiner from the Boston Police Department was prepared to testify that three partial prints found on the detective's car appeared conclusively to have been made at the same time, therefore by the same individual, and that he had found six points of comparison on one finger, two on another finger, and five on a third.

Even by the loose standards of American fingerprint experts regarding the minimum number of points required to declare a match, this combining of different fingers with just a few points of comparison on each one—that is, the use of "simultaneous impressions"—is quite a stretch. Although at least one of the amicus briefs, authored by a blue ribbon team of statisticians, scientists, and legal scholars, asked the court to rule that *all fingerprint evidence should be excluded* from trials until its validity has been tested and its error rates determined, the court (perhaps not surprisingly) limited its ruling to the particular testimony offered.

The arguments made in *Patterson* and in several other similar cases cite recent examples of mistakes made in fingerprint identifications offered in criminal trials. One of these was the 1997 conviction of Stephan Cowans for the shooting of a Boston policeman based on a combination of eyewitness testimony and a thumbprint found on a glass mug from which the shooter drank water. After serving six years of a thirty-five-year sentence, Cowans had earned enough money in prison to pay for a DNA test of the evidence. That test exonerated him, and he was released from prison.

In another notorious case, the lawyers defending Byron Mitchell on a charge of armed robbery in 1999 questioned the reliability of his identification based on two prints lifted from the getaway car. To bolster the prosecution's arguments on admissibility of the testimony of their fingerprint expert, the FBI sent the two prints and Mitchell's exemplar to fifty-three crime labs for confirmation. This test was not nearly so stringent as the kinds of tests that scientists have proposed, involving matching between groups of fingerprint samples. Nevertheless, of the thirty-nine labs that sent back opinions, nine (23 percent) declared that Mitchell's prints were *not* a match for the prints from the getaway car. The judge rejected the defense challenge, however, and Mitchell was convicted and sent to prison. As of this writing, the FBI has not repeated this sort of test, and the bureau still claims that there has never been a case where one of their fingerprint experts had given court testimony based on an erroneous match. That claim hangs by a slender thread, however, in light of the following story.

AN FBI FINGERPRINT FIASCO: THE BRANDON MAYFIELD CASE

On the morning of March 11, 2004, a series of coordinated bombings of the commuter train system in Madrid killed 191 people and wounded more than two thousand. The attack was blamed on local Islamic extremists inspired by Al Qaeda. The attacks came three days before Spanish elections, and an angry electorate ousted the conservative government, which had backed the U.S. effort in Iraq. Throughout Europe and the world, the repercussions were enormous. No surprise, then, that the FBI was eager to help when Spanish authorities sent them a digital copy of fingerprints found on a plastic bag full of detonators discovered near the scene of one of the bombings—fingerprints that the Spanish investigators had not been able to match.

The FBI's database included the fingerprints of a thirty-seven-year-old Portland-area lawyer, Brandon Mayfield, obtained when he served as a lieutenant in the United States Army. In spite of the relatively poor quality of the digital images sent by the Spanish investigators, three examiners from the FBI's Latent Fingerprint Unit claimed to make a positive match between the crime-scene prints and those of Mayfield. Though Mayfield had never been to Spain, the FBI was understandably intrigued to find a match to his fingerprints: He had converted to Islam in the 1980s and had already attracted interest by defending a Muslim terrorist suspect, Jeffrey Battle, in a child custody case. Acting under the U.S. Patriot Act, the FBI twice surreptitiously entered his family's home and removed potential evidence, including computers, papers, copies of the Koran, and what were later described as "Spanish documents"—some homework papers of one of Mayfield's sons, as it turned out. Confident that they had someone who not only matched the criminal fingerprints but also was plausibly involved in the Madrid bombing plot, the FBI imprisoned Mayfield under the Patriot Act as a "material witness."

Mayfield was held for two weeks, then released, but he was not fully cleared of suspicion or freed from restrictions on his movements until four days later, when a federal judge dismissed the "material witness" proceedings against him, based substantially upon evidence that Spanish authorities had linked the original latent fingerprints to an Algerian. It turned out that the FBI had known *before* detaining Mayfield that the forensic science division of the Spanish National Police disagreed with the FBI experts' opinion that his fingerprints were a match for the crime-scene prints. After the judge's ruling, which ordered the FBI to return all property and personal documents seized from Mayfield's home, the bureau issued a statement apologizing to him and his family for "the hardships that this matter has caused."

A U.S. Attorney in Oregon, Karin Immergut, took pains to deny that Mayfield was targeted because of his religion or the clients he had represented. Indeed, court documents suggested that the initial error was due to an FBI supercomputer's selecting his prints from its database, and that the error was compounded by the FBI's expert analysts. As would be expected, the government conducted several investigations of this embarrassing failure of the bureau's highly respected system of fingerprint identification. According to a November 17, 2004, article in *The New York Times*, an international team of forensic experts, led by Robert B. Stacey, head of the quality-assurance unit of the FBI's Quantico, Virginia, laboratory, concluded that the two fingerprint experts asked to confirm the first expert's opinion erred because "the FBI culture discouraged fingerprint examiners from disagreeing with their superiors." So much for the myth of the dispassionate, objective scientist.

WHAT'S A POOR MATHEMATICIAN TO DO?

In TV land, Don and Charlie would not rest until they found out not only who committed the garroting murder, but whether Carl Howard was innocent of the previous crime, and, if so, who was the real killer. Predictably, within the episode's forty-two minutes (the time allotted between commercials), Charlie was able to help Don and his fellow agents apprehend the real perpetrator—*of both crimes*—who turned out to be the eyewitness who identified Carl Howard from the police lineup (a conflict of interest that does not occur too often in actual cases). The fingerprint identification of Carl Howard was just plain wrong.

Given the less than reassuring state of affairs in the real world, with the looming possibility of challenges to fingerprint identifications both in new criminal cases and in the form of appeals of old convictions, many mathematicians and statisticians, along with other scientists, would like to help. No one seriously doubts that fingerprints are an extremely valuable tool for crime investigators and prosecutors. But the principles of fairness and integrity that are part of the very foundations of the criminal justice system and the system of knowing called science demand that the long-overdue study and analysis of the reliability of fingerprint evidence be undertaken without further pointless delay. The rate of errors in expert matching of fingerprints is clearly dependent on a number of mathematically quantifiable factors, including:

- the skill of the expert
- the protocol and method used by the expert in the individualization process
- the image quality, completeness, and number of fingers in the samples to be compared
- the number of possible matches the expert is asked to consider for a suspect print
- the time available to perform the analysis
- the size and composition of the gallery of exemplars available for comparison
- the frequency of near agreement between partial or complete prints of individual fingers from different people.

Perhaps the biggest driver for consideration of such quantifiable factors in the coming years will not be the demands of the criminal justice system, but the need for substantial development and improvement of automated systems for fingerprint verification and identification—for example, in "biometric security systems" and in rapid fingerprint screening systems for use in homeland security.

FINGERPRINTS ONLINE

By the time the twentieth century was drawing to a close, the FBI's collection of fingerprints, begun in 1924, had grown to more than 200 million index cards, stored in row after row of filing cabinets (over 2,000 of them) that occupied approximately an acre of floor space at the FBI's Criminal Justice Information Services Division in Clarksburg, West Virginia. The bureau was receiving more than 30,000 requests a day for fingerprint comparisons. The need for electronic storage and automated search was clear.

The challenge was to find the most efficient way to encode digitized versions of the fingerprint images. (Digital capture of fingerprints in the first place came later, adding an extra layer of efficiency, though also raising legal concerns about the fidelity of such crucial items of evidence when the alteration of a digital image is such an easy matter.) The solution chosen made use of a relatively new branch of mathematics called wavelet theory. This choice led to the establishment of a national standard: the discrete wavelet transform-based algorithm, sometimes referred to as Wavelet/Scalar Quantization (WSQ).

Like the much more widely known JPEG-2000 digital image encoding standard, which also uses wavelet theory, WSQ is essentially a compression algorithm, which processes the original digital image to give a file that uses less storage. When scanned at 500 pixels per inch, a set of fingerprints will generate a digital file of around 10 MB. In the 1990s, when the system was being developed, that would have meant that the FBI needed a lot of electronic file space, but the problem was not so much the storing of files, but of moving them around the country (and the world) quickly, sometimes over slow modem connections to law enforcement agents in remote locations. The WSQ system reduces the file size by a factor of 20, which means that the resulting file is a mere 500 KB. There's mathematical power for you. To be sure, you lose some details in the process, but not enough to be noticeable to the human eye, even when the resulting image is blown up to several times actual fingerprint size for a visual comparison.*

The idea behind wavelet encoding (and compression) goes back to the work of the early nineteenth-century French mathematician Joseph Fourier, who showed how any real-world function that takes real numbers and produces real number values can be represented as a sum of multiples of the familiar sine and cosine functions. (See Figure 7.) Fourier himself was interested in functions that describe the way heat dissipates, but his mathematics works for a great many functions, including those that describe digital images. (From a mathematical standpoint, a digital image *is* a function, namely one that assigns to each pixel a number that represents a particular color or shade of gray.) For almost

*The FBI did consider using JPEG, but the special nature of fingerprint images—essentially narrowly separated, "black," parallel, curved lines on a "white" background—meant that it was much more efficient to use a specially tailored system. For many images, such as a fairly uniform background, JPEG-2000 can achieve a compression rate of 200.

all real-world functions you need to add together infinitely many sine and cosine functions to reproduce the function, but Fourier provided a method for doing this, in particular for computing the coefficient of each sine and cosine function term in the sum.

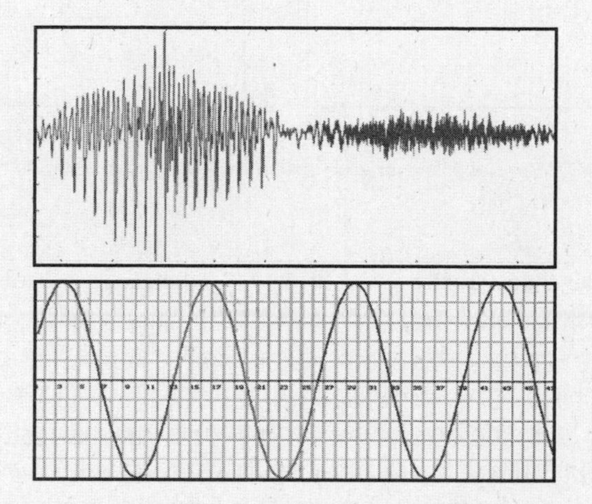

Figure 7. Fourier analysis of a wave (such as the sound wave shown above) represents it as an infinite sum of sine waves (such as the one shown below) of different frequencies and amplitudes.

Part of the complexity of Fourier analysis, and the reason it usually takes infinitely many sine and cosine terms to generate a given function, is that the sine and cosine functions continue forever, undulating in a regular wave fashion. In the 1980s, a few mathematicians began to play with the idea of carrying out Fourier's analysis using finite pieces of a wave, a so-called wavelet. (See Figure 8.) The function that generates such a wavelet is more complicated than the sine and cosine functions, but the extra complexity of the function is more than compensated by the great increase in simplicity of the resulting representation of a given function. The idea is to start with a single "mother wavelet," and create daughters by translating (shifting) the mother by one unit or else expanding or contracting it by a power of 2. You then express your given function as a sum of daughter wavelets generated by the single mother.

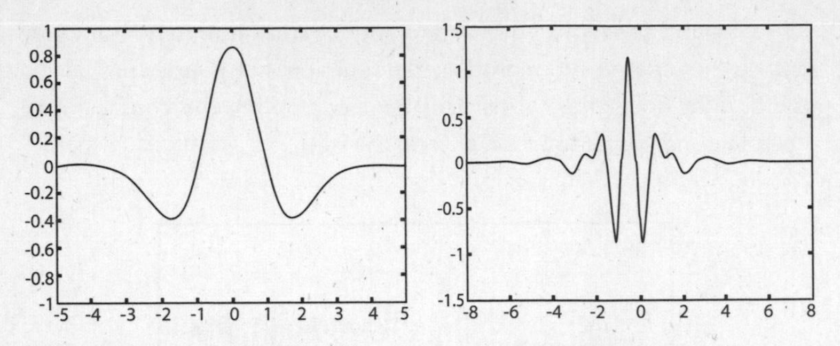

Figure 8. Wavelets. The one on the left is called, for obvious reasons, the "Mexican hat."

Wavelet theory really took off in 1987, when Ingrid Daubechies, of AT&T Bell Laboratories, constructed a family of wavelets that made this process particularly efficient when the calculations were carried out on a computer. It was not long after Daubechies' advance that the FBI started to look seriously at using wavelets to encode fingerprints. Instead of coding the bits that make up a digitized fingerprint image, the FBI's computer encodes the key numerical parameters (coefficients) in the wavelet representation of the image function. When a law enforcement agent asks for a particular set of fingerprints to be displayed on a computer screen or printed out, the computer actually *re-creates* the image using the coefficients stored in the file.

With fingerprints encoded as sequences of numbers, it becomes a relatively easy task to carry out automated computer searches looking for a match of a fingerprint in the database with one obtained from, say, a crime scene. The computer searches for strings of numbers that are very close to the string of numbers that comes from the sample. (You have to adopt a mathematically sophisticated approach to decide what "very close" amounts to in this situation; apart from that, it's a straightforward process.)

One fascinating property of wavelet encoding is that it automatically picks out the same features of an image that our eyes do. The wavelet coefficients in the final representation correspond to pixels that are very different from their neighbors, typically at the edge of the objects in the image. This means that wavelets re-create an image mostly by drawing edges—which is exactly what we do when we draw a sketch. Some researchers have suggested that the analogy between wavelet transforms and human vision is no accident, and that our neurons filter visual signals in a manner similar to wavelets.

CHAPTER

10 Connecting the Dots

The Math of Networks

PROTEST

A homemade bomb explodes beneath a car parked outside a U.S. Army recruitment office in downtown Los Angeles, killing a nearby pedestrian and injuring his wife. The bombing has all the earmarks of an anti–Vietnam War bombing carried out thirty-five years earlier to the day, even down to the message sent to the FBI (this time by e-mail) claiming responsibility and promising further attacks, the only change in the wording being the substitution of "Iraq" for "Vietnam".

The FBI had always believed the 1971 bombing was the work of an antiwar protester named Matt Stirling, who had fled immediately after the bombing and had never been apprehended. Don's first thought is that Stirling has returned to carry out some sort of anniversary repeat, and he retrieves all the files from the earlier case.

Still, it could be a copycat bombing carried out by some other person or group. But if it was, the new perpetrators would have to have had access to some detailed information about the previous event, so maybe the old case could provide clues to who organized the new one. Either way, Don has to find out all he can about the 1971 bombing. Charlie looks on as his brother works through the mountain of information.

DON: Right now, Stirling's our prime suspect. But thirty-five years is a long time to pick up a trail.

CHARLIE: But it seems you have a lot of data from the original case. I can use a branch of math called social network analysis—it looks at the structure of groups, how lines of connection develop, reveals hidden patterns. It can tell us about how Stirling fit into the organization, which in turn could tell us who he worked closely with, and the people he influenced.

DON: Your math could tell us if it's a copycat?

CHARLIE: It will identify the most likely suspects, including whether or not Stirling lands on that list.

This is how viewers of the second-season episode of *NUMB3RS* called "Protest," broadcast on March 3, 2006, were introduced to social network analysis, a relatively new branch of mathematics that became hugely important in the wake of 9/11.

A NEW KIND OF WAR, A NEW KIND OF MATH

The events of 9/11 instantly altered American perceptions of the words "terrorist" and "network", and the United States and other countries rapidly started to gear up to fight a new kind of war against a new kind of enemy. In conventional warfare, conducted in specific locations, it was important to understand the terrain in which the battles will be fought. In the war against terror, there is no specific location. As 9/11 showed only too well, the battleground can be anywhere. The terrorists' power base is not geographic; rather, they operate in networks, with members distributed across the globe. To fight such an enemy, you need to understand the new "terrain": networks—how they are constructed and how they operate.

The mathematical study of networks, known as network theory or network analysis, is based on a branch of pure mathematics called graph theory, which studies the connections between points in a set. In using techniques of graph theory and network analysis to analyze social networks, such as terrorist networks, mathematicians have developed a specialized subdiscipline known as social network analysis (SNA). SNA saw rapid development in the years leading up to 9/11 and has been an even hotter topic since. The applicability of SNA to fight crime and terrorism had been known to specialists for many years, but it was only after the Al Qaeda 9/11 plot became known that the general public realized the critical importance of "connecting the dots" in investigations and surveillance of terrorists.

THE 9/11 ATTACKS AS A CASE STUDY

The basic facts are now well known: On the morning of September 11, 2001, four commercial airliners were hijacked and turned into weapons by Al Qaeda terrorists. Two of them were crashed into the World Trade Center in New York, one into the west wing of the Pentagon in Washington, D.C., and another, believed to be heading for the White House, was heroically diverted by passengers, who perished along with the terrorists when the plane crashed in a field seventy-five miles from Pittsburgh, Pennsylvania.

The nineteen terrorists who boarded the planes that day were carrying out a plot orchestrated by Pakistan-born Khalid Sheik Mohammed, who was captured in 2003. The formal inquiry later conducted by the panel known as the 9/11 Commission outlined the information and warnings that American intelligence agencies had prior to the attacks. The Department of Homeland Security has vowed that all of the intelligence agencies would henceforth share the information needed for analysts to "connect the dots" and prevent future terrorist attack plans from succeeding.

How do mathematicians contribute to this effort? And what sort of methods do they use to analyze terrorist networks?

It is difficult to do justice to the range and power of the mathematical methods used by intelligence agencies in what has become known as the War on Terror. In fact, it's not just difficult to describe all the techniques used, it is illegal—some of the best work done by mathematicians on these problems is highly classified.

The National Security Agency, for instance, known to be the largest single employer of research-level mathematicians in the world, and affiliated organizations such as the Centers for Communications Research

(CRC), employ some of the most powerful and creative mathematical problem-solvers in the world. These mathematicians develop highly specialized methods and use them to solve real-world problems in cryptology, speech and signal processing, and counterterrorism. The NSA and similar organizations also maintain an extensive network of their own—a network of mathematicians from universities (including both authors of this book) who work with them from time to time to help develop new methods and solve hard problems. (In an early episode of *NUMB3RS*, FBI agent Don Eppes is surprised to learn that his younger brother Charlie has consulted for the NSA and has a security clearance at a higher level than Don's.)

Perhaps the best way (and the safest for your two authors) to provide a glimpse of some of the methods used is to look at studies that have been done by experts outside of the intelligence networks, using publicly available information. One of the most interesting public analyses of the 9/11 terrorists was published in April 2002 in the online journal *First Monday*. The article "Uncloaking Terrorist Networks" was written by Valdis E. Krebs, a mathematically trained management consultant with extensive experience in applying social network analysis to help clients like IBM, Boeing, and Price Waterhouse Coopers understand the way information flows and relationships operate in complex human systems. Krebs used some standard SNA calculations to analyze the structure of parts of the Al Qaeda network that (publicly available documents showed) were involved in the 9/11 attack. Figure 9 shows a graph of relationships among some of the key individuals, considered by Krebs and a later analysis published on his website (orgnet.com). The links indicate direct connections between terrorists suspected in early investigations, beginning in January 2000, when the CIA was informed that two Al Qaeda operatives, Nawaf Alhazmi and Khalid Almihdhar (shown in one of the boxes) had been photographed attending a meeting of known terrorists in Malaysia, after which they returned to Los Angeles, where they had been living since 1999. The other box contains Fahad al Quso, whose connection to Almihdhar was established when both attended the Malaysia meeting. Al Quso and Walid Ba' Attash appeared later in 2000 on the list of suspects in the October 12 bombing of the USS *Cole* while the destroyer was sitting in the Yemeni port of

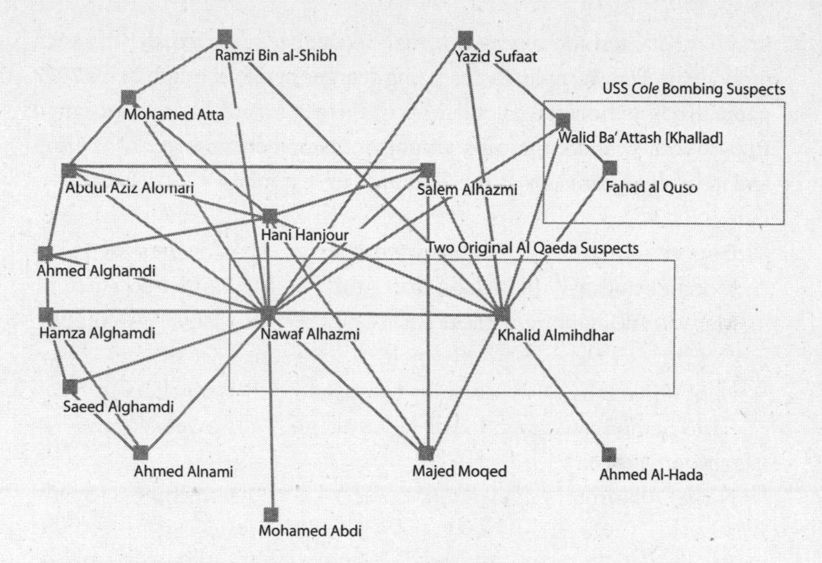

Figure 9. Graph of the Al Qaeda group behind the September 11 attacks.

Aden, an attack that killed seventeen sailors. Included in the network shown in Figure 9 are eleven of the nineteen September 11 terrorists, all of whom have either a direct link to Almihdhar and Alhazmi, the original suspects, or else are indirectly connected at a distance once removed.

Of course, this network graph was drawn "after the fact" of the 9/11 attacks and the subsequent investigations. The key challenge for investigators—and therefore for mathematicians—is to extract information in advance, typically from much larger pictures, including hundreds or even thousands of individuals. Such large networks are likely to give rise to many false leads. Normally they will also suffer from the highly troublesome phenomenon of missing data—for example, names of important participants who are absent from the graph because their existence is not known or who are present but whose links to others in the graph are not known.

A particularly important challenge is to identify in a large network those individuals who play key roles—as leaders, as facilitators, as communications "go-betweens," and so on. The mathematical tools of graph theory and social network analysis can be applied to identify such individuals. For example, in analyzing a larger network graph in his 2002 paper, Krebs performed calculations of three standard "scores" designed to point out who are the most important people in a network. The top five individuals for each of these scores were as follows.

Degree score	*Betweenness score*	*Closeness score*
Mohamed Atta	Mohamed Atta	Mohamed Atta
Marwan Al-Shehhi	Essid Sami Ben Khemais	Marwan Al-Shehhi
Hani Hanjour	Zacarias Moussaoui	Hani Hanjour
Essid Sami Ben Khemais	Nawaf Alhazmi	Nawaf Alhazmi
Nawaf Alhazmi	Hani Hanjour	Ramzi Bin al-Shibh

At the top of the list for all three calculated scores is Mohamed Atta, whose role as the ringleader of the 9/11 plot was acknowledged by Osama bin Laden in a notorious videotape released soon after the attacks. Others, such as Alhazmi, one of the two original suspects, and Hanjour and Al-Shehhi, were among the nineteen who boarded the planes on 9/11 and died that day. Others were not aboard the planes but played key roles: Moussaoui, later convicted as the "twentieth hijacker," Bin al-Shibh, Atta's roommate in Germany who couldn't gain entry to the United States, and Ben Khemais, the head of Al Qaeda's European logistical network, later convicted in Milan on conspiracy charges in another plot.

The fact that these key individuals were singled out from a network graph much larger than the one shown above, using standard social network analysis calculations, illustrates the usefulness of such calculations, which are currently being performed thousands of times a day by computer systems set up to help analysts monitor terrorist networks.

BASIC GRAPH THEORY AND "MEASURES OF CENTRALITY"

To understand the calculations used to single out the key individuals in a network graph, we need to assemble a few basic ideas. First of all, the mathematical concept of a graph as used in the present discussion is not the same as the more common notion of "graphing a curve" with vertical and horizontal axes. Rather, it refers to a set of points called nodes—people, for example—with some pairs of nodes connected by an edge and other pairs of nodes not connected. These so-called simple graphs, with no multiple edges allowed between the same two nodes, are used to represent the existence of some relationship, such as "works with" or "has a bond with" or "is known to have communicated with". Whenever two nodes do not have an edge connecting them, it means that a relationship does not exist—or is not known to exist.

Pictures of graphs are helpful, but the same graph can be represented by many different pictures, since the location of the nodes in a picture is chosen entirely as a matter of convenience (or to make things look nice). Mathematically, a graph is not a picture; it is an abstract set of nodes (also called vertices), together with edges connecting certain pairs of nodes.

A basic notion of graph theory that turns out to be important in social network analysis is the degree of a node—that is, the number of other nodes directly connected to it by edges. In a graph describing a human network, nodes of high degree represent "well-connected" people, often leaders. (Note that the word "degree" here has a different meaning from the one associated with the phrase "six degrees of separation," which is discussed later in this chapter.)

But direct connections are not all that matters. Another important notion is the "distance" between two nodes. Any two nodes are considered connected (possibly indirectly) if there is some path between them—that is, some sequence of nodes starting at one and ending at the other, with each node connected to the next by an edge. In other words, a path is a route between two nodes where one travels along edges, using intermediate nodes as "stepping-stones." The length of a path is the number of edges it contains, and the shortest possible length of a path between nodes A and B is called the distance between them, denoted by d(A,B). Paths that have this shortest possible length are called geodesic paths. In particular, every edge is a geodesic path of length 1.

The notion of distance between nodes leads to other ways of identifying key nodes—that is, it leads to other measures of centrality that can be used to give each node a "score" that reflects something about its potential importance. The concept of "betweenness" gives each node a score that reflects its role as a stepping-stone along geodesic paths between other pairs of nodes. The idea is that if a geodesic path from A to B (there may be more than one) goes through C, then C gains potential importance. More specifically, the betweenness of C as a link between A and B is defined as

the number of geodesic paths from A to B that go through C

divided by

the number of geodesic paths from A to B.

The overall betweenness score of C is calculated by adding up the results of these calculations for all possible examples of A and B. Here is an example of a node in a graph having low degree but high betweenness:

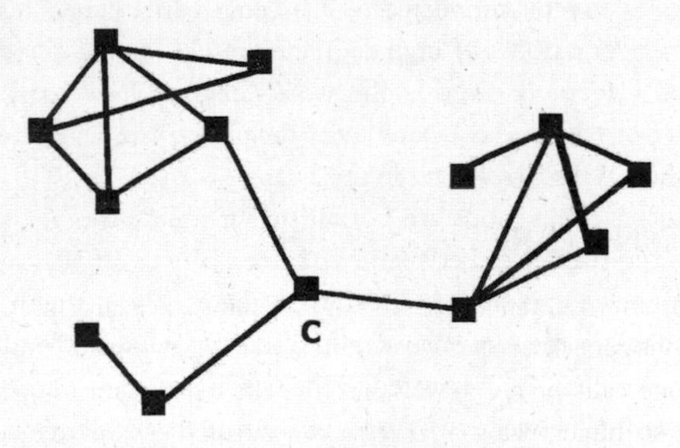

Such nodes—or the people they represent in a human network—can have important roles in providing connections between sets of nodes that otherwise have few other connections, or perhaps no other connections.

The third "centrality measure" used by Krebs, and shown in the table above, is the "closeness" score. Roughly speaking, it indicates for each node how close it is to the other nodes in the graph. For a node C, you first calculate the distances d(C,A), d(C,B), and so on, to all of the other nodes in the graph. Then you add the reciprocals of these distances—that is, you calculate the sum

$$1 / d(C,A) + 1 / d(C,B) + \ldots$$

The smaller the distances are between C and other nodes, the larger these reciprocals will be. For example, if C has 10 nodes at distance 1 (so C has degree 10), then the closeness calculation starts with 10 ones, and if there are an additional 60 nodes at distance 2, then we add "½" 60 times, and if there are 240 nodes at distance 3, then we add "⅓" 540 times, getting

$$10 \times 1 + 60 \times 1/2 + 240 \times 1/3 \ldots = 10 + 30 + 80.\ldots$$

Whereas degree measures count only immediately adjacent nodes, closeness gives credit for having many nodes at distance 2, many more at distance 3, and so on. Analysts consider closeness a good indication of how rapidly information can spread through a network from one node to others.

RANDOM GRAPHS: USEFUL TOOLS IN UNDERSTANDING LARGE NETWORKS

The amount of detailed information contained in a large graph, such as the graphs generated by the NSA in monitoring communications including phone calls or computer messages in regions such as the Middle East, is so huge that mathematicians naturally want to find "scaled-down models" for them—similar graphs that are small enough that their features can be studied and understood, and which can then provide clues about what to look for in analyzing the actual graphs. Recent research on graphs and networks has led to an explosion of interest in what are called random graphs. These graphs can help not only in understanding the structural features of large graphs and networks, but in estimating how much information is missing in a graph constructed from incomplete data. Since it is virtually impossible to get complete data about communications and relationships between people in a network—particularly a covert network—this kind of estimation is critically important.

Interest in the study of random graphs was sparked in the late 1950s by the research of two Hungarian mathematicians, Paul Erdös and Alfred Renyi. What they investigated were quite simple models of random graphs. The most important one works like this:

> Take a certain number of nodes n. Consider every pair of nodes (there are $n \times (n-1)/2$ pairs) and decide for each of these pairs whether they are connected by an edge by a random experiment—namely, flip a coin that has probability p of coming up heads, and insert an edge whenever the flip results in heads.

Thus, every edge occurs at random, and its occurrence (or not) is entirely unaffected by the presence or absence of other edges. Given its random construction you might think that there is little to say about such a graph, but the opposite turns out to be the case. Studying random graphs has proved useful, particularly in helping mathematicians understand the important structural idea called graph components. If every node in a graph has a path leading to every other node, the graph is said to be connected. Otherwise, the nodes of the graph can be separated into two or more components—sets of nodes within which any two are connected by some path, but with no paths connecting nodes belonging to different components. (This is a mathematician's way of describing the "You can't get there from here" phenomenon.)

Erdös and Renyi showed that values of p close to $1/n$ are critical in determining the size and number of components in a random graph. (Note that any one node will be connected by an edge to $(n-1) \times p$ other nodes—on average. So if p is close to $1/n$ the average degree of all the nodes is about 1.) Specifically, Erdös and Renyi demonstrated that if the number of edges is smaller than the number of nodes by some percentage, then the graph will tend to be sparsely connected—with a very large number of components—whereas if the number of edges is larger by some percentage than the number of nodes, the graph will likely contain one giant component that contains a noticeable fraction of the nodes, but the second-largest component will likely be much smaller. Refinements of these results are still a subject of interesting mathematical research.

The study of random graphs has seen an explosion of interest in the late 1990s and early 2000s on the part of both pure mathematicians and social network analysts, largely thanks to the realization that there are far more flexible and realistic probability models for the sorts of graphs seen in real-world networks.

Since real-world networks are constantly evolving and changing, the mathematical investigation of random graphs has focused on models that describe the growth of graphs. In a very influential paper written in 1999, Albert Barabasi and Reka Albert proposed a model of preferential attachment, in which new nodes are added to a graph and have a fixed quota of edges, which are randomly connected to previously existing nodes with probabilities proportional to the degrees of the existing nodes. This model achieved stunning success in describing a very important graph—namely, the graph whose nodes are websites and whose connections are links between websites. It also succeeded in providing a mechanism for generating graphs in which the frequency of nodes of different degrees follows a *power law distribution*—that is, the proportion of nodes that have degree n is roughly proportional to $1/n^3$. Later research has yielded methods of "growing" random graphs that have arbitrary powers like $n^{2.4}$ or $n^{2.7}$ in place of n^3. Such methods can be useful in modeling real-world networks.

SIX DEGREES OF SEPARATION: THE "SMALL WORLD" PHENOMENON

Another line of mathematical research that has recently attracted the attention of network analysts is referred to as the "small world model." The catalyst was a 1998 paper by Duncan Watts and Steven Strogatz, in which they showed that within a large network the introduction of a few random long-distance connections tends to dramatically reduce the diameter of the network—that is, greatest distance between nodes in the network. These "transitory shortcuts" are often present in real-world networks—in fact, Krebs' analysis of the 9/11 terrorist network described judiciously timed meetings involving representatives of distant branches of the Al Qaeda network to coordinate tasks and report progress in preparing for those attacks.

The most famous study of such a phenomenon was published by social psychologist Stanley Milgram in 1967, who suggested that if two U.S. citizens were picked at random they would turn out to be connected on average by a chain of acquaintances of length six. Milgram's basis for that claim was an experiment, in which he recruited sixty people in Omaha, Nebraska, to forward (by hand!) letters to a particular stockbroker in Massachusetts by locating intermediaries who might prove to be "a friend of a friend of a friend." In fact only three of fifty attempts reached the target, but the novelty and appeal of the experiment and the concept underlying it ensured its lasting fame.

The more substantial work of Watts and Strogatz has led to more accurate and useful research, but the "six degrees" idea has gained such a strong foothold that mythology dominates fact in popular thinking about the subject. The phrase "six degrees of separation" originated in the title of a 1991 play by John Guare, in which a woman tells her daughter, ". . . everybody on the planet is separated by only six other people. . . . I am bound, you are bound, to everyone on this planet by a trail of six people. It is a profound thought." It's not true, but it's an intriguing idea.

What does in fact seem to be true is that the diameters of networks—the longest path lengths (or average path lengths) between nodes—are smaller than one would expect based on the sheer size of the networks. There are two intriguing examples that are much talked about in widely separated fields. In the movie business, the "Kevin Bacon game" concerns the connections between film actors. Using actors as nodes of a graph, consider two actors as connected by an edge if they have appeared in at least one movie together. Because the actor Kevin Bacon has appeared in movies with a great many other actors, the idea originated some years ago to show that two actors are not far apart in this graph if both have a small "Bacon number," defined as their geodesic distance from Kevin Bacon. Thus, an actor who appeared in a movie with Kevin would have a Bacon number of 1, and an actor who never appeared with him but was in a movie with someone whose Bacon Number is 1 would have a Bacon number of 2, and so on. A recent study yielded the following distribution of Bacon numbers:

0	1	2	3	4	5	6	7	8
1	1,673	130,851	349,031	84,615	6,718	788	107	11

The average distance from Kevin Bacon for all actors in the study was 2.94. Accordingly, a conservative estimate of the distance between any two actors (obtained by adding their distances from Kevin Bacon) yields about 2 times 2.94—approximately 6! Of course, this is conservative (Kevin Bacon may not be on the shortest path between two actors), but it also falls short of satisfying "six degrees of separation" for the actors-in-the-same-movie graph, since some actors already have a distance from Kevin Bacon that is greater than 6. (Of course, actors know many other actors they haven't appeared in a movie with.)

Mathematicians have a different hero—the same Paul Erdös we met earlier. Erdös was one of the most prolific mathematicians of the twentieth century, writing more than 1,500 papers with more than 500 coauthors. In 2000, using data from sixty years of mathematical papers in research journals, Jerrold Grossman constructed a "mathematical collaboration graph" with 337,454 nodes (authors) and 496,489 edges connecting authors who wrote at least one paper together. The average degree is 3.92 and indeed there is one "giant component" containing 208,200 vertices, with the remaining 45,139 vertices contained in 16,883 components. The "Erdös number" of a mathematician is the shortest distance from that mathematician to Paul Erdös. By convention it is 0 for Erdös himself, 1 for the 500-plus mathematicians who wrote papers with him, 2 for those who wrote at least one paper with an Erdös coauthor, and so on. (Both authors of this book have an Erdös number of 2; Devlin actually wrote a paper with Erdös once, but it was never published, so it doesn't count.) At the time of Grossman's study, the average Erdös number for all published mathematicians was 4.7. The largest known Erdös number is 15.

AN EXAMPLE OF SUCCESSFULLY CONNECTING THE DOTS

One of the goals of social network analysis is to estimate which edges are missing in a graph constructed from incomplete information. For example, the "triad problem" concerns the phenomenon of "triangularity." If A, B, and C are three nodes of a network, and it is known that a certain relationship exists between A and B and also between A and C, then there is some likelihood that the same relationship—perhaps "knows" or "communicates with" or "works with"—exists between B and C also. Such likelihoods are best expressed as probabilities, and mathematicians try to determine how to estimate those probabilities based on all of the information available. For particular kinds of networks and relationships, detailed information about the connection between A and B and the connection between A and C can be used to make intelligent guesses about the probability of a relationship between B and C. Those guesses can be combined with other sources of information about a network in a way that enhances the ability of an analyst to identify the key nodes that deserve the greatest attention in further surveillance.

On June 7, 2006, during a meeting in an isolated safehouse near Baqubah, Iraq, Abu Musab al-Zarqawi, the leader of Al Qaeda in Iraq and the most-wanted terrorist in that war zone, was killed by bombs dropped by American F-16 fighter jets. Locating and killing al-Zarqawi, who had led a vicious terrorist campaign that included the capture and televised beheadings of American civilians working in Iraq, had been for several years an extremely high-priority goal of the governments of the United States, Iraq, and Jordan. Accordingly, considerable effort and manpower were devoted to tracking him down.

Although details of the methods used are closely guarded secrets, it is known that the movements and communications of a large network of al-Zarqawi's associates were monitored as closely as possible over a long period of time. One of those associates, Sheik Abdul Rahman, described as al-Zarqawi's "spiritual advisor," was pinpointed and ultimately provided the critical link. As U.S. military spokesman Major General William Caldwell said,

> Through a painstaking intelligence effort, we were able to start tracking him [Abdul Rahman], monitor his movements, and establish when he was doing his linkup with al-Zarqawi. . . . It truly was a very long, painstaking, deliberate exploitation of intelligence, information gathering, human sources, electronics, and signal intelligence that was done over a period of time—many, many weeks.

One can only imagine what the network graphs constructed by U.S. intelligence analysts looked like, but evidently the key step was identifying and zeroing in on a node at distance 1 from the most important target.

CHAPTER

11 The Prisoner's Dilemma, Risk Analysis, and Counterterrorism

In the first season of *NUMB3RS*, an episode called "Dirty Bomb," broadcast on April 22, 2005, highlighted a very real, and scary, terrorism scenario: the threatened detonation of a "dirty bomb," where radioactive material is packed around conventional explosive, with the intention that the detonation will spread deadly radioactive material over a wide area. In the episode, a team of domestic terrorists hijacks a truck carrying canisters of cesium 137, a radioactive isotope. A breakthrough in the FBI's investigation leads to a raid on the criminals' hideout, and three members of the team are taken into custody. Unfortunately, the truck, the radioactive material, and at least one coconspirator remain at large, and the men in custody brazenly threaten that if they are not released, the bomb they claim to have assembled will be set off in Los Angeles.

Don and his FBI colleagues use conventional interrogation methods, separating the three suspects and trying to get each of them to reveal the location of the truck in return for a plea bargain deal. But the three have another idea: Release them first, and *then* they'll reveal where to find the truck. Don seeks Charlie's help to resolve the stalemate.

Charlie sees a way to use a classic mathematics problem, the "prisoner's dilemma," from the branch of mathematics called game theory. Charlie explains the problem in its standard form, involving just two prisoners:

> Say two people were to commit a crime. If neither talks, they each get a year. If one talks, he gets no time, the other does five years. If both talk, they both get two years.

A possible rationale for the scenario is this: If only one of the prisoners talks, he will go free as a reward for his promise to testify at the trial of the other prisoner, who will receive the full five-year sentence upon conviction. If neither talks, successful prosecution will be more difficult, and the defense lawyers will plea-bargain for one-year sentences. If both prisoners talk, they will both get a sentence of two years, rather than five, for their cooperation, which avoids a trial.*

This scenario poses a major dilemma. The worst overall outcome for both prisoners is to talk; if they do, they both get two years. So it would seem sensible for each to stay quiet, and serve a year. But if you were one of the prisoners, having reasoned that it is better to stay quiet and serve one year, why not change your mind at the last moment and rat on your partner, thereby getting off scot-free? Seems like a smart move, right? In fact, it would be dumb not to do that. The trouble is, your partner will surely reason likewise, and the result is you both end up spending two years in prison. The more you try to puzzle it out, the more you find yourself going around in circles. In the end, you have to give up, resigned to having no alternative than to pursue the very action that you both know leads to a worse outcome.

If you are still unconvinced that your dilemma is truly hopeless, read on. Like Charlie, we'll look at the problem mathematically and derive a concrete answer.

HOW MATHEMATICIANS DEFINE A GAME

The theory of games became a mathematical discipline with the publication in 1944 of the book *The Theory of Games and Economic Behavior* by John von Neumann and Oskar Morgenstern. Their way of

*It turns out that the actual numbers—one year, two years, five years—are not important, just the *comparisons between them*, but we'll stick with the figures we have.

defining the game Charlie is describing is in terms of a payoff matrix, like this:

		Prisoner #2's Strategy	
		Trust	Talk
Prisoner #1's Strategy	Trust	Both get 1 year	#1 gets 5 years
	Talk	#2 gets 5 years	Both get 2 years

Note that in each case where one prisoner talks and the other doesn't, the one who talks goes free while his trusting partner gets five years.

Now let's see if we can figure out what is the best strategy for prisoner #1. (The analysis for #2 is exactly the same.)

A strategy is called "dominated" if it yields worse results than another strategy no matter what the other player does. If one strategy is dominated, then the other strategy would have to be a better choice—right? Let's see.

If you are prisoner #1, you always do better by talking than trusting. Assuming your partner talks, you get two years rather than five years; assuming your partner trusts you, you go free rather than get one year. So "trust" is a dominated strategy, and "talk" is a better choice for you—no matter what the other does! (Game theory assumes that players are both rational and selfish, and that the payoff matrix is the whole story. So unless the payoffs incorporate "cost of selling out my fellow prisoner" in some way—which they could—the reasoning just given is airtight.)

But wait, there's more. Notice that if both prisoners use the best strategy, the result is that both serve two years, whereas if they both used the inferior strategy, "trust," the result is actually better—both serve only one year. Aha! So what is best for the players individually is *not* best for them collectively. The phenomenon that game theorists call cooperation is at work here. *If* the prisoners cooperate with each other, and trust each other not to talk, *then* they will get the best possible outcome.

This seeming paradox—the conflict between rational self-interest and what can be achieved through cooperation—had a powerful influence on the development of game theory in the second half of the twentieth century. The prisoner's dilemma itself was first proposed by two mathematicians, Merrill Flood and Melvin Dresher, at the RAND Corporation, a government think tank that pioneered the application of mathematical methods to U.S. government strategy. Game theory was an important tool for military strategists during the cold war, and as we shall see, it is still important in mathematical analyses of strategies in the war against terrorism.

The mathematician John Nash, whose mathematical brilliance and struggle with mental illness were both dramatized in the award-winning film *A Beautiful Mind,* won a Nobel Prize in Economics for the breakthrough in game theory he achieved while earning his Ph.D. in mathematics at Princeton University. His theory, concerning what are now called Nash equilibria, is about "unregrettable" strategies—that is, combinations of strategy choices by individual players that no player can ever regret and say, "I could have done better if I'd used strategy X instead." For any game with two or more players, each having a finite list of possible strategies, Nash proved that there will be at least one such equilibrium—at least one combination of strategies for the players that is stable in the sense that no player can obtain a higher payoff by changing strategy if no one else changes.

Nash's idea was that in a game in which all players are rational and selfish, trying only to maximize the payoff to themselves, the only possible stable outcomes are these equilibria, since all other combinations of strategy choices by the players will offer at least one player a potentially greater payoff by changing strategy. Often these equilibria involve what game theorists call "mixed strategies," in which each player is allowed to use more than one of the strategies in their list (the so-called pure strategies), provided they assign a probability to each one and select a pure strategy at random according to those probabilities. In the battle of wits ("game of strategy") between a pitcher and a batter in baseball, for example, the pitcher might choose among the pure strategies of fastball, curveball, and change-up, with probabilities of 60 percent, 33 percent, and 7 percent in order to keep the batter guessing.

For the payoff matrix shown above for the prisoner's dilemma, there is only one combination of strategies that yields a Nash equilibrium, and that is a combination of two pure strategies—both prisoners choose "talk." If either prisoner departs from that strategy without the other changing strategy, then that departure causes an increase in their sentence, from two years to five. But if *both* change strategies, then they both improve their payoff, reducing their sentences from two years to one.

PLAY IT AGAIN, SAM

Prisoner's dilemma and other similar paradoxes helped spur the development of more general mathematical formulations, such as the notion of two players repeatedly playing the same game, which offers the possibility that the players will learn to trust each other by improving their payoffs. This leads to interesting possibilities, and in a famous experiment conducted around 1980, Robert Axelrod, a mathematical political scientist at the University of Michigan, organized a tournament by inviting colleagues around the world to write computer programs that would play sequences of prisoner's dilemma games against each other without any communication of intentions or "deal-making." Each entrant's program could rely only on how its opponent's program was playing the game.

The winner of the prisoner's dilemma tournament was determined by simply keeping score: What was the average payoff won by each program against all other programs? The surprising winner was a program called "Tit for Tat," written by Anatol Rapoport. The simplest of all of the programs entered, it behaved according to the following rule: Choose "trust" in the first game, and in later games choose whatever strategy the other player chose in the game before. This program is neither too nice—it will immediately punish the other player for choosing "talk"—nor too aggressive, since it will cooperate as long as the other player is cooperating. Even without the luxury of communication between the players, the Tit for Tat strategy seems to attract other computerized "players" to play the same way that it plays, leading to the best possible outcome for both.

In the fictitious scenario depicted in *NUMB3RS*' "Dirty Bomb" episode, clearly there *was* prior communication among the three criminals, and evidently they agreed to tough it out if apprehended, believing that this attitude would force the FBI to release them in order to prevent a radiation catastrophe. Similar departures from the usual assumptions of game theory are used in ongoing efforts by mathematicians to analyze and predict the strategies of terrorists and to determine the best strategies to defend against them. One of the ways to apply other mathematical ideas to enhance game theory is actually the same method that Charlie used to break apart the team of criminals, which we look at next.

RISK ASSESSMENT

The idea behind risk assessment (sometimes called "risk analysis" or "risk management") is that an individual or group confronted with possible losses can assign numerical values to those losses—perhaps actual dollar costs—and, by considering for each loss both its cost and its probability of occurring, determine the expected loss or risk it represents. They can then consider courses of action that reduce the risks, though the actions might incur some costs, too. The overall goal is to find the best combination of actions to minimize the overall cost—the cost of the actions plus the risks remaining after the actions are taken.

Among the earliest applications of risk assessment were the calculations made by insurance companies to determine how much money they should expect to pay in claims each year and the probability that the total claims will exceed financial reserves. Likewise, many companies and government agencies perform mathematical assessments of risks of various kinds, including natural disasters such as catastrophic accidents, fires, floods, and earthquakes, and take actions such as buying insurance and installing safety equipment to reduce those risks in a cost-effective manner.

Risk assessments can be made in the criminal justice system, too, and they are routinely made by defendants, their lawyers, and prosecutors, albeit usually without the benefit of actual mathematics. What Charlie realizes when confronted with the FBI's version of the prisoner's dilemma—how to crack the solidarity of the "nobody talks" strategy of the criminals in custody—is that their shared strategy subjects the three of them to very unequal risks. When Don laments that none of them show any willingness to talk, Charlie responds, "Maybe that's because none of them realizes how much the others have to lose."

Charlic convinces Don to try a different approach: Bring the three men into one room and give them a mathematical assessment of their individual risks (in the game-theoretic sense) in going to prison. Since each of them has—in one way or another—a non-negligible probability of going to prison for their participation in the dirty bomb plot, Charlie wants to show them how different the consequences would be for them individually.

Although Charlie is intimidated by facing these men—a group not at all like his usual audience of eager CalSci students—he bravely goes ahead, mumbling, "What I'm going to do today, mathematically, is construct a risk assessment for each of you. Basically quantify, if I can, the various choices you face and their respective consequences."

Gaining confidence, he writes on the board the numbers that describe their individual circumstances, saying, "Now I'll need to assign some variables, based on things like your respective ages, criminal records, loved ones on the outside . . ."

Over the heated objections of the ringleader, whom Charlie has labeled "G" on the blackboard, the lecture comes to a conclusion.

"Okay, there it is. Fitchman, you have a risk assessment of 14.9. 'W', you have 26.4, and 'G', you have a risk assessment of, oh, 7.9."

Fitchman asks, "What does that mean?" and Don replies, "It means that Ben here ['W' on the board] has the most to lose by going to prison."

Don and Charlie elaborate, talking about Ben's youth, his lack of a criminal record, his close family ties, and so on, leading to Charlie's summary of his risk assessment for the young man: "Therefore, as I've shown mathematically, you have the most to lose if you don't cooperate."

What follows is undoubtedly the first "math-induced copping of a plea" in the history of television! Far-fetched? Perhaps. But Charlie's math was spot on.

REAL-WORLD RISK ASSESSMENT VERSUS TERRORISM

These days, many mathematical tools are brought to bear on the problem of combating terrorism—data mining, signal processing, analysis of fingerprints and voiceprints, probability and statistics, and more. Since the strategies of both terrorists and defenders involve considerations of what the other side will do, the application of game theory is an attractive option, much as it was throughout the cold war. But as we saw in the case of the prisoner's dilemma and the fictional "Dirty Bomb" episode on *NUMB3RS*, there are limitations to game theory as a means of determining the best courses of action. The use of side communications and the formation of agreements among players, the uncertainties about which strategies they are actually using—what game theorists call "incomplete information"—and the difficulty of determining realistic payoffs as judged by the players, all combine to make the challenges facing game theorists extremely difficult.

Risk assessment is a key ingredient in mathematicians' efforts to supplement or even replace game-theoretic analyses. A good example is given in the recent (2002) paper "Combining Game Theory and Risk Analysis in Counterterrorism: A Smallpox Example"* by David L. Banks and Steven Anderson.

Their analysis of the threat of a smallpox attack by terrorists uses the scenarios that many government experts and other researchers have focused upon. These comprise three categories of possible attacks:

- no smallpox attack
- a lone terrorist attack on a small area (like the infamous post-9/11 anthrax letters in the United States)
- a coordinated terrorist attack on more than one city

and four scenarios for defense:

---

*In *Statistical Methods in Counterterrorism*, Alyson G. Wilson, Gregory D. Wilson, David H. Olwell, editors (New York: Springer, 2006).

- stockpile smallpox vaccine
- stockpile vaccine and develop biosurveillance capabilities
- stockpile vaccine, develop biosurveillance, and inoculate key personnel
- vaccinate everyone in advance (except the "immunocompromised")

Banks and Anderson consider the game-theoretic payoff matrix for the three attack strategies versus four defense strategies as essentially twelve boxes to be filled in, each one containing the dollar cost (or equivalent) to the defender. To determine the numerical values to put in those boxes, they propose using a separate risk assessment for each box. For example, the combination of strategies ("no smallpox attack", "stockpile vaccine") incurs a cost that the authors describe (as of the June 2002 government decision-making) as

$$ET_{\text{Dry}} + ET_{\text{Avent}} + ET_{\text{Acamb}} + VIG + PHIS,$$

where

$ET_{\text{Dry}}, ET_{\text{Avent}}$ = costs of efficacy and safety tests for the Dryvax and Aventis vaccines,

ET_{Acamb} = cost of new vaccine production and testing from Acambis,

VIG = cost of sufficient doses of Vaccinia Immune Globulin to test adverse reactions,

PHIS = cost of setting up the public health infrastructure to manage the stockpiling.

At the time of the authors' analysis, a government contract fixed the Acambis cost at $512 million, but the costs for testing Dryvax and Aventis vaccines involve clinical trials and possible follow-ups.

Moreover, there is great uncertainty about the cost of production and testing of sufficient doses of VIG and about the costs of the PHIS, the public health infrastructure. The key to the authors' mathematical analysis is to derive estimates of these uncertain dollar amounts from expert opinions. Rather than use a single best guess of each amount, they propose using ranges of plausible values, expressed by probability distributions. For example, they model the public health infrastructure's cost as the familiar bell-shaped curve, centered at $940 million, with a spread (a standard deviation) of $100 million.

Once the risk assessments for the twelve possible combinations of attack/defense strategies are made, Banks and Anderson see how the game plays out by sampling possible payoff matrices—with definite numbers filled in—using the probability distributions that describe the experts' opinions. It is essentially like drawing out of a hat possible answers to all of the unanswered questions, generating different payoff matrices one after another, each of which could be true. For each payoff matrix, they calculate a performance score for each of the four defense strategies. These scores describe the cost incurred by each defense strategy when the attacker uses their best possible strategy (a maximin strategy, in game-theory lingo).

Using the best expert opinions available in 2002, Banks and Anderson found in their computer simulations that the most effective strategy for defense was the "vaccinate everyone" strategy. But they caution that their results are not conclusive, since all four defense strategies scored in comparable ranges, indicating that the uncertainty in the public debates on U.S. strategy is not unreasonable. In recommending that their mathematical methods should be applied to future analyses of terrorist threats and defensive strategy, Banks and Anderson argue that using game theory and risk assessment methods *together* is better than using either approach alone. That is because risk assessment by itself fails to capture the kind of interaction between adversaries ("If he does this, I can do that") that game theory incorporates naturally, whereas game theory ordinarily requires definite payoffs rather than the probabilistic analysis of payoffs that risk assessment accommodates.

OPERATIONS RESEARCH VERSUS NUCLEAR WEAPONS IN SHIPPING CONTAINERS

Among the terrorist threats that were heavily debated during the 2004 presidential election campaigns was the possibility of smuggling of nuclear materials and weapons into the United States through seaports. It is widely believed that a system of defense against this threat should include inspections of shipping containers at overseas ports before they are loaded onto ships bound for the United States. At the world's second-busiest port, in Hong Kong, a demonstration project for such inspections was set up by the Hong Kong Container Terminal Operators Association. Inspections there are conducted as follows.

- Trucks carrying a shipping container on its way to be unloaded onto a ship must be permitted to pass through a gate.
- Seventy-five meters in front of the gate, the trucks must pass through a portal and be scanned by a radiation portal monitor (RPM) that detects neutron emissions.
- If the RPM cannot determine that the container contents pose no risk, the container can be diverted to a customs inspection facility for a different type of scan and possible physical inspection of its contents.

The Hong Kong pilot program was designed so that the trucks would pass the portal with the RPM detector at a speed of ten miles per hour, permitting a scan time of approximately three seconds. Longer scan times would permit detection of lower rates of neutron emissions, but slowing down the progress of the line would incur costs. The inspection protocol has to specify other variables, too, including the targeting of certain containers for closer scrutiny based on the automated targeting system of the U.S. Customs and Border Protection service. This is an expert system that uses the data accompanying each container shipment, its cargo manifest, along with possible intelligence information and observable indicators that suggest a container is more likely to be "dirty."

The key to the Hong Kong demonstration project is to avoid slowing the flow of trucks into the unloading area. The RPM scans have to be carried out without causing a slowdown, which would significantly increase the cost of port operations. The details of the setup include a branching of the queue after the trucks enter the front gate into four lanes, each with a guard who verifies the drivers' identities and tells them where to go to drop off their containers.

The Hong Kong system was carefully designed to be efficient. But just as Charlie Eppes is rarely satisfied with any system he hasn't had an opportunity to analyze mathematically, a group of real-world operations researchers (see below for an explanation of what that term means) decided to set up a mathematical model of every aspect of the Hong Kong system—the RPM scanning of the line of trucks before the front gate, the protocol for analyzing the scans and choosing some for further investigation, and the cost of the whole operation.

In their paper "The Optimal Spatial Deployment of Radiation Portal Monitors Can Improve Nuclear Detection at Overseas Ports" (2005), Lawrence M. Wein, Yifan Liu, Zheng Cao, and Stephen E. Flynn analyze mathematically a set of alternative designs for the nuclear screening of container shipments to determine whether it is possible to improve upon the effectiveness of the Hong Kong project's design. Before explaining their ideas, however, we should answer the question: What is operations research and how could it lead to a better-designed system?

Operations Research (OR) refers to a wide range of mathematical tools and methods that are applied to what is sometimes called "the science of better"—that is, the analysis of how real-world operations work and how to make them work better. Originally applied during the period after World War II to military systems like logistics, supply, and naval warfare, OR soon found other uses—to increase the efficiency of business operations, public facilities (including airports, amusement parks, and hospitals), public services such as police departments and paramedics, and many government operations and services. The tools in operations research are all mathematical—for example, the use of mathematical models for complex systems, algorithms, computer simulations, probability theory, and statistics. Sometimes the term "management science" is used as a rough synonym for operations research.

Applications of OR in police work have included mathematical investigations of how to distribute patrols in high-crime areas, how to guard high-profile targets, and how to organize and analyze data for use in investigations. Many universities have departments of operations research or management science, and faculty members, in addition to teaching, typically do both theoretical research on mathematical methods and consulting on real-world problems.

One of the classical components of OR is queueing theory, a branch of probability theory that investigates the phenomena associated with "waiting in line" (known in the United Kingdom as "queueing," hence the theory's name), and seeks to provide answers to questions such as "What is the most efficient way to design the waiting lines in a bank?" or "How many tellers will be needed to limit the average waiting time to five minutes if the customers flow in at a rate *r*, and each requires an average of *t* minutes to be serviced?"

Now let us return to the trucks waiting patiently in Hong Kong to unload their containers onto ships. We'll see how operations researchers like Wein, Liu, Cao, and Flynn use mathematical analysis to design better systems and calculate their performance and cost. We start with a picture of the flow of trucks and containers through the front gate:

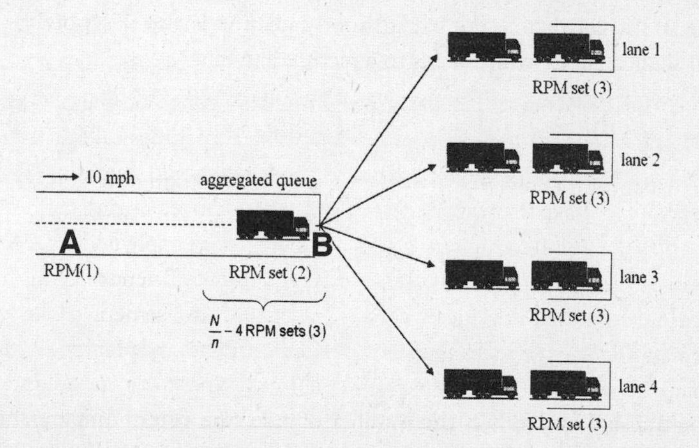

The Hong Kong demonstration experiment places an RPM at point A, 75 meters ahead of point B, the front gate. This ensures that trucks can flow past the monitor at a regulated speed of 10 miles per hour. Since each 40-foot shipping container is carried lengthwise on a truck, it takes about three seconds for it to pass through, so that the monitor collects three seconds worth of neutron emission counts. The number of neutrons counted depends on A, ε, S, τ, and r, where

A = area of the neutron detector = 0.3 square meters,
ε = efficiency of the detector = 0.14,
S = neutrons emitted per second (depending on the source),
τ = testing time = number of seconds the RPM is allowed to count the neutrons,
r = distance from the RPM to the center of the container = 2 meters.

The result is:

$$\text{average number of neutrons counted} = A\varepsilon S\tau \,/\, 4\pi r^2.$$

The variability of the number counted is described by a bell-shaped curve whose width (or standard deviation) is about 2.8 times the square root of the average. Since there is background radiation of neutrons at rate B, smaller than S, the background radiation is also described by a bell-shaped curve, which leads to a picture like this:

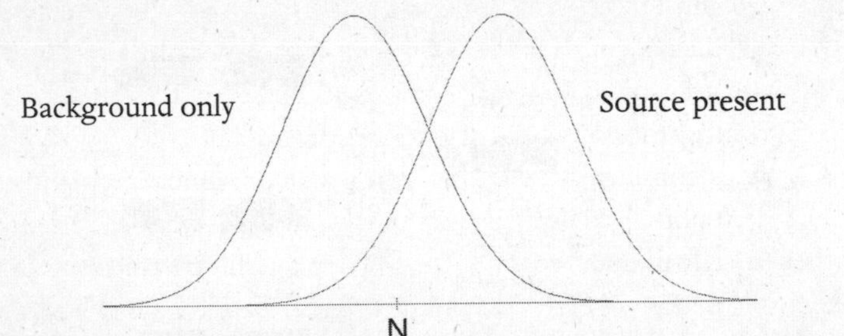

The threshold value N is the number of neutrons detected that call for another level of scrutiny—asking a human analyst to examine the scan produced by a VACIS gamma-ray-imaging system, which is designed to detect the kind of dense material in the container that would be used to shield emissions. If the person reading that scan cannot confirm the safety of the container, the truck is diverted to an offsite location where customs inspectors do a high-energy X-ray scan and, if necessary, open the container and inspect its contents manually. These inspections are a relatively costly part of the total system, but they can reliably detect radioactive material. Even if no containers exceed the RPM threshold, 5 percent of them can be expected to be flagged for VACIS inspection as untrusted containers by the automated targeting system, which uses a separate risk analysis of containers based on information about their origins.

Other key variables are the probabilities of success in the VACIS and X-ray scans and the costs include:

- $250 for each high-energy X-ray
- $1,500 for each opening of a container and manual inspection
- $100,000 for the annualized cost of each RPM machine

The objective of the entire analysis is to devise systems that for a given annual cost achieve the lowest possible detection limit:

S_D = source level of neutrons per second the RPM can detect

with the requirement that the probability of the RPM detecting that source level must be at least 95 percent. False positives—that is, containers that produce a count at the level N or higher because of naturally occurring background radiation—are considered in the model, too, since they incur the costs of additional testing.

All things considered—within a constraint on annual cost and the requirement not to slow down the flow of trucks—what can be done mathematically to improve the system? Wein and his coauthors analyze the existing design together with three possibly better ways to operate:

Design 1 = (existing) RPM at location A, 75 meters before front gate

Design 2 = RPM at the front gate, B

Design 3 = 4 RPMs, one at each lane-processing point

Design 4 = Add to Design 3 a row of 10 RPMs in the line in front of gate B

Under the quantitative assumptions of their paper, the OR mathematicians show that over a range of annual cost levels:

- Design 2 improves the detection limit S_D by a factor of 2 with the same cost.
- Design 3 improves S_D by an additional factor of 4.
- Design 4 improves S_D by an additional factor of 1.6.

Thus, the overall improvement in going from Design 1, as used in the Hong Kong experiment, to Design 4 is a factor of 13 reduction in the source level of neutron radiation that the system can detect. How is this achieved?

The answer is in two parts. First is the fact that the longer the testing time, τ, the greater the probability that we can correctly distinguish the presence of extra neutron emissions over the background. For the same reason that statisticians always recommend, if possible, taking a larger sample of data, a longer time for the RPMs to count neutrons effectively separates the bell-shaped curves so that they look more like this:

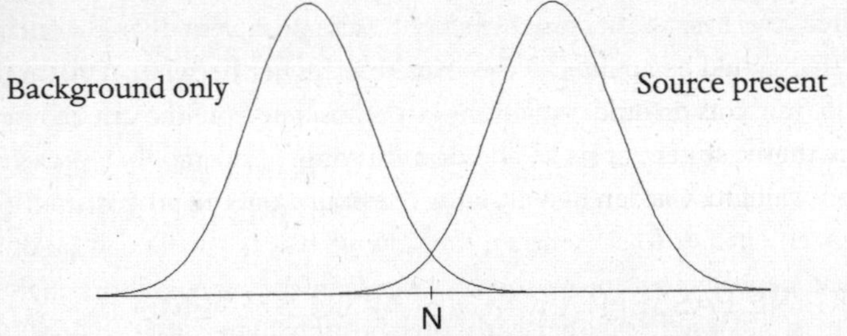

Since the two curves are now much less overlapped, the threshold value N used for detection can be set relatively lower without increasing the frequency of false positive detections. Alternatively, one can set the level of N so that false positives occur with the same frequency as before and successful detections occur when the source emissions, S, are lower. So the detection limit, S_D, is reduced.

The second part of the authors' solution comes from the analysis of the queueing models for the four designs. The goal is to expose containers to longer testing times τ. The potential for improvement is clear, since the RPM at point A gets only three seconds to look at each truck, whereas the trucks wait much longer than that to pass through the inspection process.

In moving the single RPM from A to B, Design 2 takes advantage of the fact that sometimes there are more trucks than usual flowing into the line, so that the line backs up, causing trucks to have to idle for a while in a line behind the front gate at B. So if the RPM is placed there, it will get a longer testing time on the trucks that have to wait.

In replacing the single RPM at B with four RPMs, one in each lane, Design 3 achieves an even greater improvement over Design 1, since the average processing time for the trucks to be cleared by the inspectors at the head of those lanes is sixty seconds. By using additional RPMs in a row before the front gate, B, Design 4 adds additional testing time, making possible a further reduction in the detection limit.

But what about the cost of all of those extra RPMs? Within any fixed annual budget, that cost can be offset by decreasing the frequency of false positives at each stage of the screening process, thereby reducing the cost of X-ray scans and manual inspections. The essence of an OR type of mathematical modeling and optimization (a mathematician's word for finding the best way) is that one has to determine which variables in a system should be adjusted to get better performance, while maintaining the constraints on other variables like the cost and the flow rate of trucks through the system. If he knew about the work of operations researchers like Wein, Liu, Cao, and Flynn, Charlie Eppes would be proud.

AIRLINE PASSENGER SCREENING SYSTEMS

Ever since the tragic events of September 11, 2001, the U.S. government has invested major financial and human resources in preventing such attacks from succeeding ever again. That attack intensified the government's efforts to enhance airline security through a system that had already been in place since 1998. Called CAPPS (for "computer assisted passenger prescreening system"), it relies on a passenger name record containing basic information obtained by the airline when a passenger books a ticket—name, address, method of paying for the ticket, and so on. The airline uses that information to make a check against the Transportation Security Administration's "no-fly list" of known or suspected terrorists and also to calculate a "risk score" based on terrorist profiles. These are lists of characteristics typical of terrorists, derived from statistical analysis of many years' worth of data on the flying habits of known terrorists. If a hit occurs on the no-fly list, or if the profile-based risk score is high enough, the airline subjects the passenger and his or her luggage to a more intensive, "second-level" screening than ordinary passengers undergo.

A similar system was instituted after a wave of skyjackings of commercial airliners in the years 1968 and 1969 (when there were more than fifty such events) led to the development of a "skyjacker profile" that was used for several years and then discontinued. Though the specific elements of both the skyjacker profile and the terrorist profile are closely guarded secrets, a few of their features have frequently been surmised in public discussions. (For instance, if you are a young man traveling alone, it would be better not to buy a one-way ticket, particularly if you pay for it with cash.)

After 9/11, the newly formed Transportation Security Administration assumed responsibility not only for a "no-fly list" but for the statistical analyses needed to design more effective profiles of terrorists. Experts outside the government believe that the TSA is using neural nets (see Chapter 3) to refine the terrorist profile. There is no doubt that it seems like good common sense for federal authorities to try to separate out from the general population those airline passengers who could be considered high risk as potential terrorists, and then subject them to greater scrutiny and search. That is the logic of CAPPS. But how well can such a system be expected to work? The answer, as we shall see, is not as simple as it might at first appear.

TWO MIT STUDENTS USE MATHEMATICS TO ANALYZE CAPPS

In May 2002, a pair of graduate students at MIT made national news by announcing a paper they had prepared for a class called "Ethics and Law in the Electronic Frontier." Samidh Chakrabarti and Aaron Strauss thought that analyzing CAPPS would make an interesting paper for the class, and the results of their mathematical analysis were so striking that the professor urged them to publish them more widely, which they proceeded to do on the Internet. Their paper "Carnival Booth: An Algorithm for Defeating the Computer-Assisted Passenger Screening System" caused a sensation because it showed by clear logic and mathematical analysis how terrorists could rather easily adapt their behavior to render CAPPS less effective than pure random selection of passengers for second-level screening.

The two authors assume that

- No matter which system is used to select passengers for second-level screening, only 8 percent of them can be handled in that way.
- In CAPPS, the federal requirement to randomly select "x percent of passengers" for second-level screening is met by randomly selecting 2 percent.
- Three out of every four terrorists assigned to secondary screening will be successfully intercepted.
- If they are not assigned to secondary screening, only one out of every four terrorists will be successfully intercepted.
- The percentage of terrorists who are not randomly selected for secondary screening but will be flagged by CAPPS is unknown. Call it p percent.

These percentage assumptions made by Chakrabarti and Strauss are not random. Rather, they based their analysis on the publicly available best estimates of the actual percentages, which are a government secret. Their results do not depend substantially on the exact values of those percentages. The unknown percentage p depends on how high a risk score is required to assign a passenger to secondary screening. To meet the requirement of "no more than 8 percent" for secondary screening, the threshold for the risk score will have to be chosen so that it is achieved by 6 percent of the nonterrorist passengers who escape random selection.

Then, the overall percentage of terrorists who will be intercepted under CAPPS is:

(*) 3/4 of p% + 3/4 of 2% + 1/4 of the remaining percentage

For comparison, Chakrabarti and Strauss consider a "pure chance" system, in which the 8 percent of passengers that can be handled in secondary screening are chosen at random from the list of all passengers. In that case the overall percentage of terrorists who will be intercepted is:

(**) 3/4 of 8% + 1/4 of 92% = 6% + 23% = 29%

Comparing (*) and (**), the obvious question is "Which method intercepts a higher percentage of terrorists?" The answer depends on the value of p, the unknown percentage of terrorists who are selected because they meet the profile. Here are some examples:

Value of p	Overall percentage of terrorists intercepted
2%	27%
4%	28%
6%	29%
8%	30%
10%	31%

From these examples it is clear that the break-even point for CAPPS versus a purely random system is when $p = 6$ percent of terrorists are subjected to secondary screening because they meet the profile.

Now comes the heart of the matter. You might say, "Surely we can expect the percentage of terrorists who meet the profile to be larger than a paltry 6 percent!" That is where the phenomenon Chakrabarti and Strauss call the "carnival booth effect" comes in. They argue that, since the terrorist profile is fixed, and terrorist cells have members with a diversity of characteristics, a cell that wants to be successful in getting one of its members aboard a plane for an attack can use the following strategy:

- Probe the CAPPS system by sending cell members on "dry runs" to see which ones are flagged by the profile and which are not.
- For the actual attack mission, use members who were not flagged in the dry runs and are therefore very unlikely to be flagged by the same profile next time.

Chakrabarti and Strauss call this carnival booth effect because it is reminiscent of the barkers at carnival booths who call out "Step right up and see if you're a winner!" The would-be attackers who constitute a real threat are the "winners" who do not trigger secondary screening when they "step right up" to the CAPPS profiling system.

As the MIT authors explain at some length, the viability of this strategy depends on just two essential factors: the observation that the CAPPS profile itself is fixed over time—at least over short time intervals—which implies the "repeatability" of an individual's not being selected by the profile; and the recognition that terrorist cells have members with considerable diversity of characteristics, and so are likely to include at least one member who can pass the profile part of the screening system. In support of the latter claim, they describe some of the known terrorists from recent events, such as John Walker Lindh, the "American Taliban," a nineteen-year-old from Marin County, and Richard Reid, the British citizen with an English mother and Jamaican father who single-handedly made sure that we now all have to take off our shoes before boarding an airplane.

The two MIT researchers included in their paper some more sophisticated analyses using computer simulations incorporating some variability and uncertainty in the CAPPS profile scores of each individual terrorist. For instance, they found that repeated probes would, for some individual terrorists, increase the confidence of not being flagged to a higher level than that of a randomly chosen passenger. In that case, the CAPPS probability of intercepting an actual attack by such an individual would be worse than random.

Such is the power of mathematics, that even a couple of bright college students writing a term paper can make a significant contribution to an issue as significant as airline security.

CHAPTER

12 Mathematics in the Courtroom

Okay, so Charlie has pulled out all the mathematical stops and as a result Don has once again nailed his suspect. That is generally the end of a *NUMB3RS* episode, but in real life it is often not the end of the mathematics. Math is used not only in crime detection, but in the courtroom as well.

One example is the use of mathematically enhanced photographs, as in the Reginald Denny beating case described in Chapter 5; another is the probability calculations that must accompany the submission of DNA profile evidence, which we looked at in Chapter 7. But there are many other occasions when lawyers, judges, and juries must weigh mathematical evidence. As our first example shows, if they get the math wrong, the result can be a dramatic miscarriage of justice.

THE BLONDE WITH THE PONYTAIL

Just before noon on June 18, 1964, in the San Pedro area of Los Angeles, an elderly woman named Juanita Brooks was walking home from grocery shopping. Using a cane, she was pulling a wicker basket containing her groceries, with her purse on top. As she made her way down an alley, she stooped to pick up an empty carton, and suddenly she felt herself being pushed to the ground. Stunned by the fall, she still managed to look up, and saw a young woman with a blond ponytail running away down the alley with her purse.

Near the end of the alley, a man named John Bass was watering the grass in front of his house when he heard crying and screaming. He looked over toward the alley and saw a woman run out of it and get into a yellow car across the street. The car started up, turned around, and took off, passing within six feet of him. Bass subsequently described the driver as a "male Negro" (this was 1964) wearing a beard and a mustache. He described the young woman as Caucasian, slightly over five feet tall, with her dark blonde hair in a ponytail.

Brooks reported the robbery to Los Angeles police, telling them her purse had contained between $35 and $40. Several days later, they arrested Janet Louise Collins and her husband Malcolm Ricardo Collins, who were ultimately charged with the crime and placed on trial in front of a jury.

The prosecutor faced an interesting challenge. Neither eyewitness, Brooks or Bass, could make a positive identification of either of the defendants. (Bass had previously failed to identify Malcolm Collins in a lineup, where he appeared without the beard he admitted he had worn in the past—but not on the day of the robbery, he said.) There was some doubt caused by the witnesses' description of the ponytailed blonde's clothing as "dark," since the police had obtained testimony from people who had seen Janet Collins shortly before the robbery wearing light-colored clothing. How was the prosecutor to make the case to the jury that these two defendants were guilty of the purse snatching?

The prosecutor took a novel approach. He called an expert witness: a mathematics instructor at a state college. The expert testimony concerned probabilities and how to combine them. Specifically, the mathematician was asked to explain the product rule for determining the probability of the joint occurrence of a combination of events based on the individual probabilities of those events.

The prosecutor asked the mathematician to consider six features pertaining to the two perpetrators of the robbery:

Black man with a beard

Man with a mustache

White woman with blond hair

Woman with a ponytail

Interracial couple in a car

Yellow car

Next, the prosecutor gave the mathematician some numbers to assume as the probabilities that a randomly selected (innocent) couple would satisfy each of those descriptive elements. For example, he instructed the mathematician to assume that the male partner in a couple is a "black man with a beard" in one out of ten cases, and that the probability of a man having a mustache (in 1964) is one out of four. He then asked the expert to explain how to calculate the probability that the male partner in a couple meets *both* requirements—"black man with a beard" and "man with a mustache." The mathematician described a procedure well known to mathematicians, called the "product rule for independent events." This says that "if two events are independent, then the probability that both events occur together is obtained by multiplying their individual probabilities."

Thus, in the hypothetical case proposed by the prosecutor, if the events are indeed independent (we'll discuss later exactly what that means), then you can use the product rule to calculate the probability that an individual is a black man with a beard and has a mustache by multiplying the two given probabilities:

$$\begin{aligned}&P(\text{black man with a beard AND has a mustache})\\&\quad = P(\text{black man with a beard}) \times P(\text{has a mustache})\\&\quad = 1/10 \times 1/4 = 1/(10 \times 4) = 1/40\end{aligned}$$

The complete list of probabilities the prosecutor asked the mathematician to assume was:

Black man with a beard: 1 out of 10

Man with mustache: 1 out of 4

White woman with blond hair: 1 out of 3

Woman with a ponytail: 1 out of 10

Interracial couple in car: 1 out of 1,000

Yellow car: 1 out of 10

The prosecutor asked the mathematician to take those numbers as conservative assumptions, meaning that the actual probabilities would be at least this small and possibly smaller.

The mathematician then proceeded to explain how to combine these probabilities to come up with an overall probability that a random couple would satisfy all of the above description. Assuming independent events (more later), the mathematician testified that the correct calculation of the overall probability, let's call it PO, uses the same product rule, which means you multiply the individual probabilities to get the probability that the whole list applies to a random couple. When you do this, here is what you get:

$$
\begin{aligned}
PO &= 1/10 \times 1/4 \times 1/3 \times 1/10 \times 1/1000 \times 1/10 \\
&= 1/(10 \times 4 \times 3 \times 10 \times 1000 \times 10) \\
&= 1/12{,}000{,}000
\end{aligned}
$$

One out of 12 million!

When the prosecutor gave the various odds—1 in 10, etc.—to the mathematics expert to use in calculating the overall probability, he stated that these particular numbers were only "illustrative." But in his closing argument he asserted that they were "conservative estimates," and therefore "the chances of anyone else besides these defendants being there, . . . having every similarity . . . , is something like one in a billion."

The jury found Malcolm and Janet Collins guilty as charged. But did they make the right decision? Was the mathematician's calculation correct? Was the prosecutor's closing "one in a billion" claim correct? Or had the court just been party to a huge travesty of justice? Malcolm Collins said it was the latter, and appealed his conviction.

In 1968 the Supreme Court of the State of California handed down a decision in *People v. Collins, 68 Cal.2d 319*, and their written opinion has become a classic in the study of legal evidence. Generations of law students have studied the case as an example of the use of mathematics in the courtroom.

Here is what the California Supreme Court's opinion (affirmed by a six-to-one vote of the justices) said:

> We deal here with the novel question whether evidence of mathematical probability has been properly introduced and used by the prosecution in a criminal case. . . . Mathematics, a veritable sorcerer in our computerized society, while assisting the trier of fact in the search for truth, must not cast a spell over him. We conclude that on the record before us defendant should not have had his guilt determined by the odds and that he is entitled to a new trial. We reverse the judgment. . . .

The majority opinion in the Collins case is a fascinating example of the interplay between two scholarly disciplines: law and mathematics. Indeed, the majority opinion took pains to say that they found "no inherent incompatibility between the [two] disciplines" and that they intended "no disparagement" of mathematics as "an auxiliary in the fact-finding process" of the law. Nevertheless, the court ruled that they could not uphold the way mathematics was used in the Collins case.

The Supreme Court's devastating deconstruction of the prosecution's "trial by mathematics" had three major elements:

- Proper use of "math as evidence" versus improper use ("math as sorcery")
- Failure to prove that the mathematical argument used actually applies to the case at hand
- A major logical fallacy in the prosecutor's "one in a billion" claim about the chances of the defendants being innocent

Let's see just what went wrong with the prosecution's case.

MATHEMATICS: EVIDENCE OR SORCERY?

The law recognizes two principal ways in which an expert's testimony can provide admissible evidence. An expert can testify as to their own knowledge of relevant facts, or they can respond to hypothetical questions based on valid data that has already been presented in evidence. So, for example, an expert could testify about the percentages—in Los Angeles, say—of cars that are yellow, or of women who are blondes, provided there exists statistical data to support that testimony. And a mathematician can respond to hypothetical questions such as "How would you combine these probabilities to determine an overall probability?"—provided those hypotheticals are based on valid data. In the Collins case, however, the Supreme Court found that, the prosecution "made no attempt to offer any such evidence" of valid probabilities.

Moreover, the court pointed out that the prosecution's mathematical argument rested on the assumption that the witnesses' descriptions were 100 percent correct in all particulars and that no disguises (such as a false beard) were employed by the true perpetrators of the crime. (The trial record contained disputes about light versus dark clothing worn by the young woman, and about whether or not the defendant had a beard.)

The court pointed out that it is traditionally the function of juries to weigh the reliability of witness descriptions, the possibility of disguise by the perpetrators, and the like. But these considerations are not ones that can be assigned numerical probabilities or likelihoods. Moreover, the Supreme Court believed that the appeal of the "mathematical conclusion" of odds of 1 in 12 million was likely to be too dazzling in its apparent "scientific accuracy" to be discounted appropriately in the usual weighing of the reliability of the evidence. The court wrote: "Confronted with an equation which purports to yield a numerical index of probable guilt, few juries could resist the temptation to accord disproportionate weight to that index. . . ." That is at the heart of the "sorcery" that the Supreme Court found in the prosecution's case.

WAS THE COURT'S MATH CORRECT?

Leaving aside the question of whether it was permissible to use mathematics in the way the original court allowed, there is the issue of whether the math itself was correct. Even if the prosecution's choice of numbers for the probabilities of individual features—black man with a beard, and so on—were supported by actual evidence and were 100 percent accurate, the calculation that the prosecutor asked the mathematician to do depends on a crucial assumption: that in the general population these features occur *independently*. If this assumption is true, then it is mathematically valid and sensible to use the product rule to calculate the probability that the couple who committed the crime, if they were *not* Malcolm and Janet Collins, would by sheer chance happen to match the Collinses in each of these factors.

That crucial assumption of independence means that if we think of the individual probabilities as representing fractions of the general population, then those fractions continue to apply in sequence as we look at each fraction in turn. Let's consider an example that is similar and slightly easier to work with. Suppose that witnesses to a crime said that the perpetrator drove a black Honda Civic that was "lowered"—fitted with special springs that make the body sit closer to the ground.

Ignoring the likely possibility that witnesses might also identify other features of the perpetrator, let's assume that we know, accurately and based on solid data, that in the Los Angeles area 1 out of 150 cars is a black Honda Civic and 1 out of 200 is lowered. The product rule says that to determine the fraction of cars that are black Honda Civics that have been lowered, we multiply:

$$1/150 \times 1/200 = 1/30{,}000.$$

But this calculation is based on the assumption that the fraction of cars that have been lowered is the same for black Honda Civics as it is for all other makes and colors. If that were true, we could say that the descriptive features "black Honda Civic" and "lowered" occur independently. There is, however, the possibility that owners of black Honda Civics are more likely than owners of most other cars to have them customized by lowering. The correct calculation of the probability that a car in L.A. is a black Honda Civic that's been lowered (assuming we have good data to determine these numbers) is as follows.

Suppose that

fraction of cars that are black Honda Civics = 1/150

fraction of black Honda Civics that are lowered = 1/8

Then, the fraction of cars that are "lowered" black Honda Civics is

$$1/150 \times 1/8 = 1/(150 \times 8) = 1/1200$$

which is considerably larger than 1/30,000.

The number used for illustration here, 1/8, is called the "conditional probability" that a car is lowered, given that it is a black Honda Civic. Obtaining reliable data to determine that number, or at least to estimate it accurately, is likely to be more difficult than just estimating the fraction of *all* cars that have been lowered—the "1 out of 200" in the original calculation. But surely, in any serious endeavor—in particular, a criminal trial—the fact that a number is hard to determine or estimate is no excuse for making a highly dubious assumption, such as independence. The potential for error is compounded when we pile up a whole list of features (six in the Collins case) and assume that they are all independent. Even Charlie Eppes would be hard-pressed to come up with the right data and calculate an accurate estimate of the probability that a couple who commit a crime in Los Angeles would have those six characteristics.

Yet that was not the last of the errors the original trial court made. The most devastating blow that the Supreme Court struck in its reversal of Collins' conviction concerned a mistake that (like the unjustified assumption of independence) occurs frequently in the application of probability and statistics to criminal trials. That mistake is usually called "the prosecutor's fallacy."

This notorious fallacy consists of a sort of bait-and-switch tactic by the prosecution, sometimes made because of unintentional error. On the one hand, we have the prosecution's calculation, which in spite of its lack of justification, attempts to determine

P(match) = the probability that a random couple would possess the distinctive features in question (bearded black man, with a mustache, etc.)

Ignoring the defects of the calculation, and assuming for the sake of argument that P(match) truly is equal to 1 in 12 million, there is nevertheless a profound difference between P(match) and

P(innocence) = the probability that the Collinses are innocent.

As the Supreme Court noted, the prosecutor in the Collins case argued to the jury that the 1 in 12 million calculation applied to P(innocence). He suggested that "there could be but one chance in 12 million that defendants were innocent and that another equally distinctive couple actually committed the robbery."

The confusion between these two probabilities is wrong and dangerous! P(match) is trying to calculate the probability that the defendants, *if they are innocent,* would be unlucky enough to match the witness descriptions of the couple who committed the robbery. But as the justices explained in their opinion, a "probability of innocence" calculation (even if one could presume to actually calculate such a thing) has to take into account how many other couples in the Los Angeles area also have these six characteristics. The court said, "Of the admittedly few such couples, which one, if any, was guilty of committing the robbery?"

In a master stroke that warmed the hearts of mathematicians and statisticians around the world when we subsequently read about the Collins case, the court's opinion went on to add an appendix in which they calculated another estimate. Even taking the prosecution's 1 in 12 million result at face value, what is the probability that somewhere in the Los Angeles area there are at least two couples who have the six characteristics as the witnesses described for the robbers? The justices estimated that probability by assuming that there are a large number N of possible perpetrators—pairs of people (not necessarily "couples") in the Los Angeles area—and that each pair has a probability of 1 in 12 million of fitting the robbers' descriptions. Using their own independence assumption about different pairs fitting the description (which is not exactly right but is not a source of substantial error), they performed a calculation using the binomial distribution.

Imagine flipping N coins, they reasoned, each with probability 1 in 12 million of turning up heads. Given that at least one coin turns up heads (meaning that there is at least one couple that meets the description), what is the probability that two or more heads occur—that there are at least two couples that meet the description?

The answer to the question is easy to calculate using the binomial distribution (a calculator or spreadsheet can be used), and not surprisingly it depends on N—the number of potential "perpetrator couples". For illustration, the court used N = 12 million, approximately the number of people in the Los Angeles area at the time, and they calculated that the answer is "over 40 percent." (It's actually 41.8 percent.) In this way, they argued that it is not at all reasonable to conclude that the defendants must be guilty simply because they have the six characteristics in the witnesses' descriptions.

Of course, a different choice of N would give a different answer, but even N = 3 million, say, would yield a probability of 12 percent that somewhere in Los Angeles there exists at least one other pair who arguably would be as good candidates for conviction as the Collinses—at least in terms of the "proof by mathematics" that the prosecution relied on to sway the jury. That hardly sounds like "beyond a reasonable doubt," does it?

The key fact the prosecutor's fallacy overlooks is that there are typically many other people (or couples) not on trial who have the calculated probability (like 1 in 12 million) of matching the accused person (or couple). Therefore, even if those on trial are innocent, there is typically a far larger probability than P(match) of their being unlucky enough to match the characteristics being used to identify the perpetrators of the crime.

The Collins case may have become a famous example in legal circles, but it was by no means the first time in U.S. legal history that a trial was decided almost entirely on mathematics. In the Collins case, the use made of mathematics turned out to be incorrect. But things came out very differently in an equally famous case a hundred years earlier.

FAMOUS NINETEENTH-CENTURY MATHEMATICIANS DEMONSTRATE A FORGERY

One of the most famous American forgery cases, a cause célèbre in the nineteenth century, hinged upon key testimony of father-and-son mathematicians. Benjamin Peirce was one of the leading mathematicians of his day, a famous professor at Harvard University, whose name is still used to bestow honor upon young mathematicians who are appointed Benjamin Peirce Assistant Professors at Harvard. His son, Charles Sanders Peirce, was also a brilliant scholar who taught mathematical logic, worked for the U.S. Coast and Geodetic Survey, the leading federal agency in the funding of nineteenth-century scientific research, and wrote prodigiously in the field of philosophy, becoming best known as the founder of "American pragmatism."

What kind of trial would bring both of the Peirces into the courtroom as expert witnesses? It was a forgery trial involving the estate of Sylvia Ann Howland, valued at $2 million when she died—a huge figure back in 1865. Her niece, Hetty Howland Robinson, contested the will, which left her only a part of the estate, and claimed that she and her aunt had a secret agreement under which Robinson would inherit the entire estate. As proof she presented an earlier version of the aunt's will that not only left the entire estate to her but also contained a second page declaring that any later wills should be considered invalid! The executor of the estate, Thomas Mandell, rejected Robinson's claim on the basis that the second page was a forgery, and therefore the later will should determine the disposition of the estate.

Robinson was never charged with the crime of forgery. In fact, the sensational case that ensued, *Robinson v. Mandell*, popularly known as the Howland will case, resulted from Robinson's filing of a lawsuit in an attempt to overturn the executor's rejection of her claim! And this was the lawsuit that was decided using mathematics.

In most forgery cases, someone attempts to duplicate the signature or handwriting of person X, and prosecutors (or civil litigators) try to demonstrate in court the dissimilarity of the forgeries from samples of the authentic writing of X. But in the Howland will case the issue was the reverse: The forgery was simply too good!

Benjamin and Charles Peirce were called as witnesses for the defendant Mandell to testify about their careful scientific investigation of the similarity between the authentic signature on the first page and the disputed signature on a second page. (There were actually *two* second pages, but only one was analyzed.)

Here are the two signatures.

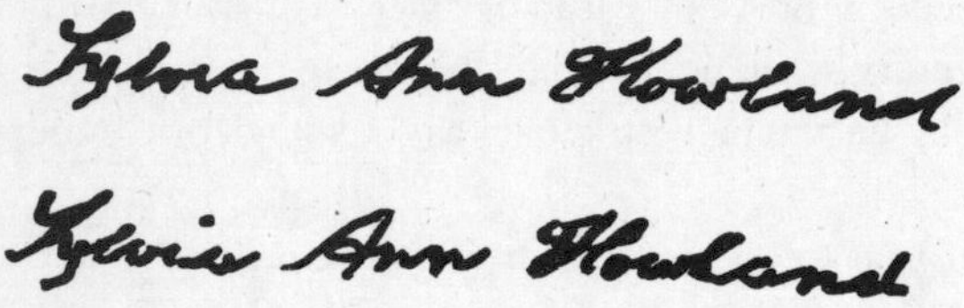

If you look at two copies of your own signature you will soon notice some differences between them. The two signatures on the Howland will, however, look identical. The most likely explanation is that one is a traced copy of the other.

What the Peirces did was turn this suspicion into a scientific fact. They devised a method to compare and express the agreement between any two signatures of the aunt as a number—a sort of score for closeness of agreement. To determine this score, they decided to use downstrokes—there are thirty of them in each signature—and to count the number of "coincidences" between the thirty downstrokes in one signature and the corresponding thirty downstrokes in the other. By a "coincidence" between two examples of a particular downstroke, such as the downstroke in the first letter "L", they mean an essentially perfect match between those strokes, which they judged by overlaying photographs of the signatures, one on top of the other.

When they compared the two signatures shown above, they found that every one of the thirty downstrokes coincided! Could this be due to sheer chance? Or was it clear evidence that the disputed signature was obtained by tracing the authentic signature onto the disputed second page? That's where the mathematical analysis came in.

The Peirces obtained a set of forty-two undisputed authentic signatures of Sylvia Ann Howland. For forty-two signatures there are $42 \times 41/2 = 861$ ways to select a pair of signatures to compare. For each of these 861 pairs they determined the number of coincidences—how many of the thirty downstrokes coincided? They found a total of 5,325 coincidences among the $861 \times 30 = 25{,}830$ comparisons of downstrokes. That meant that about one out of five comparisons was judged a coincidence—a perfect match.

The rest of their analysis was mathematical, or more specifically, statistical. The elder Peirce described his calculation of the chances of getting thirty coincidences out of thirty downstrokes, assuming that each occurred with probability $5325/25830 = 0.206156$. Assuming these coincidences occur independently (!), Peirce used the product rule to multiply, giving

$$.206156 \times .206156 \times .206156 \times \ldots \text{[30 times]}$$

i.e.,

$$.206156^{30}.$$

This figure is approximately 1 in 375 trillion. (Peirce actually made a mistake in his calculation, and gave a somewhat larger number, using 2,666 in place of 375.)

Summoning the full eloquence expected of a gentlemanly mathematician in 1868, Professor Peirce summarized his findings in this way: "So vast improbability is practically an impossibility. Such evanescent shadows of probability cannot belong to actual life. . . . The coincidence which has occurred here must have had its origin in an intention to produce it."

Surely not surprising in light of this mathematical and rhetorical splendor, the court ruled against Hetty Robinson.

What would a modern mathematician—or statistician—say about Professor Peirce's analysis? The data for the 861 comparisons of pairs of signatures—counting the number of coincidences—can be analyzed to see how well the independence assumption is satisfied, or the binomial model that it leads to, and the result is that the data counting coincidences for those 861 pairs do *not* fit Peirce's model very well at all. But that does not mean that his conclusion that the thirty coincidences on thirty downstrokes is highly unusual cannot be sustained. As pointed out by Michael O. Finkelstein and Bruce Levin in discussing the Howland Will Case in their excellent book *Statistics for Lawyers,* statisticians nowadays would typically prefer to analyze such data in a "nonparametric" way. This means the analysis does not assume that when two signatures are compared, the probabilities of zero coincidences, one coincidence, two coincidences, and on up to thirty coincidences, are known to satisfy some particular formulas or, if expressed in a bar chart, to have some particular shape.

Rather, a present-day statistician would prefer to rely on a more justifiable analysis, such as the one that says that if the null hypothesis is true (i.e., the disputed signature is authentic), then there are forty-three true signatures and thus $43 \times 42/2 = 903$ pairs of signatures, each pair with a presumably equal chance of having the greatest agreement. So, without considering how extreme thirty out of thirty is—just the fact that it shows the highest level of agreement between any of the 903 pairs of signatures—there is at most one chance out of 903 of those two particular signatures being *more alike* than any of the other pairs. Therefore, either a very unusual event has occurred—one whose probability is about one-tenth of one percent—or else the hypothesis that the disputed signature is authentic is false. That would surely be sufficient for Charlie Eppes to urge his brother to put the cuffs on Hetty Robinson!

USING MATHEMATICS IN JURY SELECTION

We suspect that few of our readers are criminals. And we certainly hope that you are not a victim of a crime. So most of the techniques described in this book will be things you merely read about—or see when you watch *NUMB3RS*. But there is a fair chance—about one in five, to be precise, if you are a U.S. citizen—that at least once in your life you will find yourself called for jury duty.

For many people, serving on a jury is the only direct experience of the legal system they experience firsthand. If this does happen to you, then there is a slight chance that part of the evidence you will have to consider is mathematical. Much more likely, however, if the case is a serious one, is that you yourself may unknowingly be the target of some mathematics: the mathematics of jury selection. This is where statisticians appointed by the prosecution or defense, or both—increasingly these days those statisticians may use commercially developed juror-profiling software systems as well—will try to determine whether you have any biases that may prompt them to have you removed from the jury.

The popular conception of a jury is a panel of twelve citizens, but in fact jury sizes vary from state to state, with federal court juries different again, from a low of six to a high of twelve. Although juries as small as three have been proposed, the general consensus seems to be that six is the absolute minimum to ensure an acceptable level of fairness.

Mathematics gets into the modern jury scene at the beginning of the selection process, as the 1968 federal Jury Selection and Service Act mandates "random selection of juror names from the voter lists." (Although the act legally applies only to federal courts, it is generally taken to set the standard.) As Charlie Eppes would tell you, randomness is a tricky concept that requires some mathematical sophistication to handle properly.

One of the goals of the jury system is that juries constitute, as far as possible, a representative cross section of society. Therefore, it is important that the selection process—which, like any selection process, is open to abuse—does not unfairly discriminate against one or more particular sectors, such as minorities. But as with the issue of determining racial bias in policing (discussed in Chapter 2), it can be a tricky matter to identify discrimination, and cases that on the surface look like clear instances of discrimination sometimes turn out to be nothing of the kind.

In one frequently cited case that went to the Supreme Court, *Castaneda v. Partida* (1977), a Mexican-American named Rodrigo Partida was indicted and convicted for burglary with intent to rape in a southern Texas border county (Hidalgo County). He appealed this conviction on the grounds that the Texas system for impaneling grand jurors discriminated against Mexican-Americans. According to census data and court records, over an eleven-year period only 39 percent of people summoned to grand jury duty had Spanish surnames, whereas 79 percent of the general population had Spanish surnames. The Supreme Court held that this was sufficient to establish a prima facie case of discrimination.

The court made its determination based on a statistical analysis. The analysis assumed that if the jurors were drawn randomly from the general population, the number of Mexican-Americans in the sample could be modeled by a normal distribution. Since 79.1 percent of the population was Mexican-American, the expected number of Mexican-Americans among the 870 people summoned to serve as grand jurors over the eleven-year period was approximately 688. In fact, only 339 served. The standard deviation for this distribution worked out to be approximately twelve, which meant that the observed data showed a difference from the expected value of roughly twenty-nine standard deviations. Since a difference of two or three standard deviations is generally regarded as statistically significant, the figures in this case were overwhelming. The probability of such a substantial departure from the expected value, often referred to as the "p value", occurring by chance was less than 1 in 10^{140}.

Another high-profile case was the 1968 district court conviction of the famous pediatrician Dr. Benjamin Spock, for advocating the destruction of draft cards during the Vietnam War. There were concerns over this conviction when it became known that the supposedly randomly selected pool of 100 people from which the jury was drawn in this case contained only nine women. According to public opinion polls at the time, antiwar sentiment was much more prevalent among women than men. Dr. Spock's defense team commissioned statistician (and professor of law) Hans Zeisel to analyze the selection of jury pools. Zeisel looked at the forty-six jury pools for trials before the seven judges in the district court in the two-and-a-half-year period before the Spock trial, and found that one judge, the one in the Spock case, consistently had far fewer women on his juror pools than any of the others. The p value for the discrepancy in this case was approximately 1 in 10^{18}. As it turned out, this clear case of discrimination was not pivotal in Dr. Spock's successful appeal, which was granted on the basis of the First Amendment.

What both cases demonstrate is how the application of a thorough statistical analysis can determine discrimination in jury selection, to a degree well beyond the standard "reasonable doubt" threshold.

However, selection of a representative jury pool is only part of the story. The American legal system allows for individual jurors to be eliminated from the pool at the beginning of the trial on three grounds.

The first ground is undue hardship on the juror. Typically, this occurs when a trial is likely to last a long time, and may involve sequestration. In such a case, mothers of small children, owners of small businesses, among others, can usually claim release from jury service. This leads many observers to the not unreasonable conclusion that lengthy trials generally have juries largely made up of people with lots of time on their hands, such as retired persons or those with independent means.

The second ground for exclusion is when one of the protagonists can demonstrate to the court's satisfaction that a particular juror is incapable of being impartial in that particular trial.

The third ground is the one that may result in a potential juror being subjected to a detailed statistical and psychological profile. This is the so-called peremptory challenge, where both prosecution and defense are allowed to have a certain number of jurors dismissed without having to give any reason. Of course, when a lawyer asks for a juror to be removed, he or she always does have a reason—they suspect that this particular juror would not be sympathetic to their case. How do they find that out?

JURY PROFILING

Although the right of peremptory challenge does give both sides in a case some freedom to try to shape the jury to their advantage, it does not give them the right to discriminate against any protected group, such as minorities. In the 1986 case *Batson v. Kentucky*, the jury convicted James Batson, an African-American, of burglary and receipt of stolen goods. In that case, the prosecutor used his peremptory challenges to remove all four African-Americans, leaving the case with an all-white jury. The case ended up in the Supreme Court, which, based on the composition of the jury, reversed the conviction. By then, Batson was serving a twenty-year sentence. Rather than risk a retrial, he pled guilty to burglary and received a five-year prison sentence.

As always, the challenge is to establish discrimination, as opposed to the effects of chance fluctuations. In another case, *United States v. Jordan,* the government peremptorily struck three of seven African-American jurors compared with three of twenty-one whites. That meant that an African-American in the jury pool was three times more likely to be excluded as a white. The p value in this case worked out to be 0.14; in other words, such a jury profile would occur by chance roughly one in every seven occasions. The court of appeal ruled that there was insufficient evidence of discrimination.

It turns out, however, that even when illegal discrimination is ruled out, prosecutors and defenders have considerable scope to try to shape a jury to their advantage. The trick is to determine in advance what characteristics give reliable indications of the way a particular juror may vote. How do you determine those characteristics? By conducting a survey and using statistics to analyze the results.

The idea was first tested in the early 1970s by sociologists enlisted in the defense of the so-called "Harrisburg Seven," antiwar activists who were on trial for an alleged conspiracy to destroy Selective Service System records and kidnap Secretary of State Henry Kissinger. The defense based its jury selection on locally collected survey data, systematically striking the Harrisburg citizens least likely to sympathize with dissidents. Far from the "hanging jury" that many observers expected from this highly conservative Pennsylvania city, the jury deadlocked on the serious charges and convicted the activists of only one minor offense.

CHAPTER

13 Crime in the Casino

Using Math to Beat the System

DOUBLE DOWN

The dealer at the blackjack table is good at her job. She jokes with the players as she deals the hands, knowing that this will encourage them to continue placing larger and larger bets. A young man sporting a goatee, long hair, and a black leather jacket comes to the table and takes an empty seat. He converts five thousand dollars into chips, and places an enormous bet on the next hand. The dealer and the other players are taken aback by the size of his bet, but the young man breaks the tension by making some remarks about his family in Moscow. He wins the hand, making a huge profit, but then, instead of playing another hand, he scoops up his chips and leaves the table. Looking for his car in the casino parking lot, he seems anxious, even afraid. Moments later he is shot and killed by an unseen assailant.

This was the opening sequence in the second-season *NUMB3RS* episode "Double Down," broadcast on January 13, 2006. As is often the case with *NUMB3RS*, the story isn't just about the crime itself, it's about the special worlds inhabited by the victims and the suspects—in this case, the world of professional blackjack players who challenge the gambling casinos. As "Double Down" unfolded, viewers learned that the victim, Yuri Chernov, was a brilliant mathematics student at Huntington Tech, making the case a natural one for Charlie to help with. To do that, he has to delve into the workings of the real-life battle of wits—and sometimes more—that has been going in the casino blackjack world for more than forty-five years.

On one side of this battle are secretive, stealthy "card counters," often working in teams, who apply sophisticated mathematics and highly developed skills in their efforts to extract large winnings from casinos. On the other side are the casinos, who regard card counters as cheaters, and who maintain files of photographs of known counters. The casino bosses instruct their dealers and other employees to be always on the lookout for new faces in the ranks of players who can walk away with tens of thousands of dollars in winnings in a matter of hours.

In most states,* players who count cards while playing blackjack are not criminals in the literal sense. But the casinos view them as criminal adversaries—cheaters, no different from the players who manipulate the chips or conspire with crooked dealers to steal a casino's money. And because of the risk of being recognized and barred from play, card counters have to act like criminals, using disguises, putting on elaborate performances to fool dealers about their true capabilities, or sneaking around, trying desperately not to be noticed.

The root cause of the casinos' difficulty is that when it comes to blackjack, an astute and suitably knowledgeable player, unlike other casino gamblers, can actually have an edge over the casino. Casinos make a profit—a generous one—by knowing the exact probabilities of winning in each game they offer, and setting the odds so that they have a slight advantage over the players, typically around 2 to 3 percent. This guarantees that, although one or two players will make a killing every now and then, the vast majority of players will not, and on a weekly or monthly basis, the casino will earn a steady profit.

In the game of craps, for example, short of actual criminal acts of cheating (manipulating chips, using loaded dice, and the like), no player can win in the long run. When honest players win, they are simply fore-

*Nevada is an exception. The lucrative gambling business in this otherwise fairly poor state enabled the casinos to exert pressure on the legislature to make card counting illegal.

stalling the losses they will eventually rack up if they come back . . . and back . . . and back. The mathematics guarantees that this will happen.

But blackjack is different. Under certain circumstances, the players have an edge. A player who can recognize when this is the case and knows how to take advantage of it can, if allowed to continue playing, capitalize on that percentage advantage to win big money. The longer the counters are allowed to play, the more they can be expected to win.

THE PROBLEM WITH BLACKJACK

In casino blackjack, each player at the table plays individually against the dealer. Both player and dealer start with a two-card hand and take turns at having the option to draw additional cards ("hit" their hand) one at a time. The aim is to get as high a total as possible (with face cards counting as 10, aces as 1 or 11), without "going bust," that is, exceeding 21. If the player ends up with a total higher than that of the dealer, the player wins; if the dealer has the higher total, the player loses. For most plays, the payoff is even, so the player either loses the initial stake or doubles it.

The twist that turned out to be a major headache for casinos is that, in the version of the game they offer, the dealer must play according to a rigid strategy. If the dealer's hand shows a total of 17 or more, he or she must "stand" (they are not permitted to take another card); otherwise the dealer is free to hit or stand.* That operational rule opens a small crack in the otherwise impregnable mathematical wall that protects the casinos from losing money.

The possibility of taking advantage of the potentially favorable rules of casino blackjack was known and exploited by only a few people until the publication in 1962 of the book *Beat the Dealer*, written by Edward Thorp, a young mathematics professor. In some ways not unlike Charlie Eppes—though without an older brother asking him to help the FBI solve crimes—Thorp was beginning his career as a research mathematician, moving from UCLA to MIT (and later to the University of California at Irvine), when he read a short article about blackjack in a mathematics journal and devel-

*Some casinos use a so-called "soft 17 rule" that requires the dealer to hit when his or her total of 17 includes an ace counted as "11".

oped an interest in the intriguing difference between blackjack and other casino games:

> What happens in one round of play may influence both what happens later in that round and in succeeding rounds. Blackjack, therefore, might be exempt from the mathematical law which forbids favorable gambling systems.*

It turns out that there are several features of the game of blackjack that are asymmetrical in their effect on the player and the dealer, not just the dealer's "17-rule." The player gets to see the dealer's first card (the so-called "up card") and can take that information into account in deciding whether to hit or "stand"—that is, the player can use a variable strategy against the dealer's fixed strategy. There are other differences, too. One asymmetry very definitely in the casino's favor is that, if both the player and the dealer bust, the dealer wins. But there are asymmetries that favor the player. For instance, the player is given the opportunity to make special plays called "doubling down" and "splitting pairs," which are sometimes advantageous. And, particularly juicy, the player gets a bonus in the form of a 3:2 payoff (rather than just "even money") when his initial two-card hand is a "natural"—an ace and a ten (picture card or "10")—unless the dealer has a natural, too.

Players can capitalize on these asymmetries because, in blackjack, after each hand is played, those cards are discarded. That means that, as the plays progress, the distribution of ten-value cards in the deck can change—something an astute player can take advantage of.

When Thorp published his revolutionary discoveries in 1962, the net effect of these asymmetries and other fine points was that the version of blackjack being played on the Las Vegas "strip" was essentially an even game, with very close to a zero advantage for the casino.

In an industry where the casinos had been used to having a guaranteed edge, Thorp's discovery was completely unexpected and impressive enough to make news, and it led to hordes of gamblers flocking to the

*Edward O. Thorp, *Beat the Dealer: A Winning Strategy for the Game of Twenty-One*, Random House, New York, 1962.

blackjack tables to play Thorp's recommended strategy, which required the player to memorize certain rules for when to hit, when to stand, and so on, depending on the dealer's up-card. All of these rules were based upon solid mathematics—probability calculations that, for example, analyzed whether a player should hit when his hand totals 16 and the dealer shows an ace. Calculating the probabilities of the various totals that the dealer might end up with and the probabilities of the totals the player would get by hitting, Thorp simply compared the probability of winning both ways—hitting and standing—and instructed players to take the better of the two options—in this case, hitting the 16.

The casinos were pleased to see the increased level of business, and they quickly realized that most of these newly minted blackjack enthusiasts only played Thorp's strategy in their dreams. Many a would-be winner had difficulty remembering the finer points of strategy well enough to execute them at the right time, or even showed a lack of dedication to the mathematically derived best strategy when subjected to the harsh realities of the luck of the draw. A run of good or bad hands—perhaps losing several times in a row by following one of the basic strategy's instructions—would often persuade players to disregard Thorp's meticulously calculated imperatives.

Nevertheless, *Beat the Dealer* was a stunning success. It sold more than 700,000 copies and made *The New York Times'* bestseller list. The game of blackjack would never be the same again.

CARD COUNTING: A MATHEMATICIAN'S SECRET WEAPON

Thorp's basic strategy, the one he developed first, simply turned a profit-maker for casinos into a fair game. How did blackjack become a potential loser for the casinos and a profit-maker for mathematicians and their avid students? Thorp carefully analyzed blackjack strategy further, using some of the most powerful computers available in the early 1960s, and he exploited two simple ideas.

One idea is for the player to vary his strategy even more (when to hit or stand, whether to double down, etc.) according to the proportion of tens left in the deck. When the chances of busting are higher than usual—say, when lots of 10s and picture cards (both counted as 10) remain in the deck and the player has a poor hand, like 16 against a dealer's 10—he can intelligently revise the basic strategy by standing instead of hitting. (If there are a lot of 10-value cards left in the deck, the chances are higher that hitting on 16 will lead to a bust.) On the other hand, when the chances of busting are lower than usual—when there are relatively more low cards in the deck—players can hit in situations where they would normally stand according to the basic strategy. These changes shift the percentage advantage from zero to a small advantage for the player.

The other idea is for the player to vary the amount bet on successive hands according to the same information—the proportion of 10-valued cards remaining in the deck. Why do that? Because the proportion of 10-value cards affects the player's prospects on the next hand. For example, if there are lots of "tens" remaining in the deck, then the chances of getting a natural go up. Of course, the dealer's chance of getting a natural goes up too, but the player gets a payout bonus for a natural and the dealer doesn't. Therefore, more frequent naturals for the player and the dealer mean a net advantage for the player!

Things would have been bad enough for the casinos if Thorp had simply explained the mathematics to the readers of his bestseller. That would have put them at the mercy of players with enough mathematical ability to understand his analysis. But Thorp did more than that. He showed them how they could count cards—that is, keep a running count of tens versus non-tens as the deck is played out—to give them a useful indicator of whether the next hand would be more favorable than average or less favorable, and to what extent.

As a result, thousands of readers of Thorp's book used its instructions to become card counters using his "Tens Strategy," and copies of the book began to appear in the hands of passengers on trains, planes, and buses arriving in Las Vegas and other parts of Nevada, where large amounts of money could be won by applying the fruits of Thorp's mathematical analysis.

The casinos were in trouble, and they immediately changed the rules of blackjack, removing certain features of the game that contributed to the player's potential to win. They also introduced the use of multiple decks shuffled together—often four, six, or even eight decks—and dealt the cards out of a "shoe," a wooden or plastic box designed to hold the shuffled cards and show the back of the next card before it is pulled out by the dealer.

Called the "perfesser stopper" in homage to Professor Thorp, whose personal winning exploits, while not huge, were sufficient to add to the enormous appeal of his book, the multideck shoes had two effects. They enabled the casinos to shuffle the cards less frequently, so that without slowing down the game (bad for profits) they could make sure to reshuffle when a substantial number of cards were still left in the shoe. This kept the card counters from exploiting the most advantageous situations, which tend to occur when there are relatively few cards left to be dealt. Moreover, the multiple-deck game automatically shifted the basic player versus house percentage about one-half of one percent in the house's favor (mainly due to the asymmetries mentioned above). Even better for the casino, dealing from multiple decks shuffled together meant that it would typically take much longer for the Thorp counting procedure to detect an advantageous deck, and the longer it took the more likely a player was to make a mistake in the count.

There was a predictable outcry from regular blackjack players about the rules changes—but only about reducing the opportunities to make plays like "doubling down" and "splitting pairs." So the casinos relented and reinstituted what essentially were the previous rules. But they kept the shoes, albeit with a few blackjack tables still offering single-deck games.

LORDEN'S STORY

At this point, we can't resist recounting the experience one of us (Lorden) had with the Thorp system.

In the summer of 1963 I was on vacation from graduate school, back home in southern California working for an aerospace company. I was fascinated by Thorp's book, particularly the part where Thorp explained how the "gambler's ruin problem" sheds light on the very practical issues of winning at blackjack. I was familiar with the problem from my mathematics studies as a Caltech undergraduate, but I had not heard of the Kelly gambling system or the other money management rules that Thorp explained.

What these rules reflect is that there is an important but little understood corollary to the well-known principle "You can't beat the odds in the long run." Many years later, in a public lecture at Caltech, I demonstrated this fact by involving the audience in an elaborate experiment.

I programmed a computer to print out 1,100 individual "gambling histories," one for each audience member, mathematically simulating the results of making steady bets on a single number on a roulette wheel, five days a week, eight hours a day, for an entire year. In spite of the 5.6 percent casino advantage at roulette, about a hundred members of the audience raised their hands when I asked, "How many of you are ahead after three months?" At the end of the lecture, the woman who got a framed certificate attesting to her performance as "best roulette player" had won a tight competition. There were three others in the audience who, like her, actually made a profit playing full-time roulette for a year! (As a seasoned presenter, before asking the computer to run the simulations and print out the results, I had calculated the probability that no one in the audience would come out a winner, and it was acceptably small.)

If random chance fluctuations can sometimes forestall for such a long time the inevitable losses in playing roulette, then perhaps it is not surprising that the flip side is also true. If I played blackjack with a winning percent advantage using Thorp's system, I still had to face the prospect of losing my meager stake before reaching the promised land of long-term winnings.

Of course, Thorp's book explained all of this and emphasized the usefulness of the Kelly gambling system, a strategy invented by a physicist at Bell Laboratories in the 1950s, which instructs that you should never bet more than a certain percentage of your current capital—typically about the same percentage as your average percentage advantage over the casino. In theory, this strategy would completely eliminate the possibility of "gambler's ruin." Unfortunately, casino games have minimum bets, so that if your capital ever gets down to, say, five dollars, betting a small percentage of it is not allowed. Playing one last hand at that point will of course give you a good chance of losing your whole stake—a genuine case of gambler's ruin.

TEAMS TAKE ON THE CASINOS

The initial response of the casinos to the success of Thorp's book turned out to be just the first round in an ongoing war between math-types and the casinos. Students of mathematics and its profitable applications quickly realized that multiple-deck blackjack, in spite of obvious disadvantages compared to single decks, has some very attractive and exploitable features. For one, it's easier to disguise card counting with multiple decks, because whenever the composition of the remaining cards becomes favorable for the player, it tends to stay favorable—perhaps for many hands. Fluctuations in the player versus dealer advantage are dampened by the presence of many cards remaining in the deck.

Also, blackjack players started playing in teams, something else that needed the much longer play cycles of multiple decks. One of the pioneers of team play was Ken Uston, who gave up his job as vice president of the Pacific Stock Exchange to devote himself full-time to winning money at blackjack. His book *Ken Uston on Blackjack* popularized methods of team play against the casinos that greatly enhanced the potential of card counters to extract profits.

In its simplest form, team play involves members pooling their money and sharing the net proceeds of their individual wins and losses. Since it can take many hands for a small percentage edge to turn into actual winnings, a team of, say, five players, playing as one, can improve their chances significantly, since they can afford to play five times as many hands than if they played individually.

Moreover, teams can avoid detection much more effectively by adopting the classic economic principle of specialization of labor. What Uston proposed was "big player teams," an idea credited to his mentor, a professional gambler named Al Francesco. Here is the idea. A casino can detect card counters because they need to change the size of their bets—suddenly changing from making small bets when the odds are on the casino's side and then placing big bets when the remaining cards are in their favor. But by playing as a team, one player can avoid detection by not betting anything unless the deck is sufficiently favorable, and then making only big bets.

The idea is for some team members to act as "spotters." Their job is to play quietly at several tables, placing small bets, all the time counting cards out of the shoe at their table. When one of them sees a favorable deck start to emerge, they signal the "big bettor" to come over to that table and take advantage of it. The big bettor thus moves around from table to table, making only large bets (and generally raking in large wins), leaving a particular table when the counters signal that the deck has turned unfavorable. The small bets made by the spotters have little effect on the team's overall winnings or losses, which come predominately from the big bettor. The main risk in this strategy is that someone watching the big bettor moving around in this fashion can recognize what's going on, but in the hustle and bustle of a large busy casino with literally dozens of blackjack tables, a skilled and experienced team can often play their well-choreographed game all night long without being detected at all.

The potential for generating steady profits from this sort of team play began to attract considerable interest among mathematics students at many universities. For most of the 1990s, teams from MIT, in particular, became highly effective in raiding gambling casinos in Nevada and other parts of the country. Their winnings were not consistent (chance fluctuations always play an inescapable role), their stealth techniques and disguises were not always effective, and their personal experiences ranged from inspiring to abysmal. But overall they gave the casinos quite a run for their money. Many of these exploits were chronicled in a popular book, *Bringing Down the House* by Ben Mezrich, in magazine and newspaper articles, in a television documentary (in which your other author, Devlin, became the only mathematician to play a James Bond role on the screen), and in the recent movie *21* (which is the alternative name for the game of blackjack).

So what's happening today in blackjack casinos? Almost certainly there are some unchronicled math-wise card counters still playing, but the casinos' countermeasures now include some high-tech machinery: automatic card shufflers. In the early 1990s a truck driver named John Breeding had the idea to replace the shoe with a machine that would not only hold multiple decks but allow played cards to be shuffled back into the deck automatically and frequently. This lead to the development of Shuffle Master machines, now visible in many casinos, which, besides relieving the dealer of the time-wasting burden of shuffling, also relieve card counters of their potential for profits. The latest versions of these machines, called CSMs (for "continuous shuffling machines"), effectively approximate "dealing from an infinite deck," a feature that makes card counting useless. In the subculture of people who play blackjack professionally, these machines are dubbed "uncomfortable shoes."

Single-deck games still exist, but in a disturbing recent trend the casinos have been transforming them into "sucker propositions" by changing the 3:2 bonus for a natural to 6:5. This shifts the advantage a whopping 1.4 percent in favor of the casino, turning the game into little more than a salutary (and possibly expensive) lesson for the sort of person who doesn't read the fine print. (And if you don't think a 1.4 percent advantage to the casino is "whopping," you should stay well away from the gaming tables!)

The "Double Down" episode of *NUMB3RS* hinged on the idea that a rogue genius mathematician was hired as a consultant to the company that manufactures the shuffling machines, and he intentionally used a poorly chosen algorithm to control the random mixing of the cards inside the machine. He then hired mathematics students and armed them with the instructions needed to decode the pattern of the cards dealt by the machine, enabling them to anticipate the sequence of cards as they came out. The writers helped themselves to a little dramatic license there, but the point is a good one. As Charlie observes: "No mathematical algorithm can generate truly random numbers." Poorly (or maliciously) designed algorithms intended to generate random numbers can indeed be exploited, whether they appear in cell phones, Internet security, or at the tables in a casino.

A FOOTNOTE: MATHEMATICIANS AND THE GAMES THEY CHOOSE TO PLAY

Thorp himself never made a huge amount of money from his casino method—apart from the royalties from his bestselling book. But he did go on to become wealthy from applying his mathematical expertise to a different game. Shortly after his stunning success in transforming blackjack, he turned his attention to the stock market, wrote a book called *Beat the Market*, and started a hedge fund to use his mathematical ideas to generate profits in stock market trading. Over a nineteen-year period, his fund showed what Wall Street calls an "annualized net return" of 15.1 percent. That's slightly better than doubling your capital every five years.

Nowadays, Wall Street and financial firms and institutions are heavily populated with "quants"—people trained as mathematicians, physicists, and the like—who have made the study of the mathematics of finance and investment into a hugely profitable enterprise.

You get the idea.

LORDEN AGAIN: CALTECH STUDENTS TAKE ON THE CASINOS

Some years ago, about a decade after Thorp's book came out, I had an experience that brought home to me just how seriously the casinos took the threat mathematics posed to their business. By then, I was back at Caltech, my alma mater, as a professor, my brief student foray into casino life long behind me. My specialty was (and remains) statistics and probability, and I would occasionally hear stories about friends of friends making killings at the casinos. I knew of the improvements in blackjack card counting that Thorp and others had made, such as the "hi-lo" count, where the player keeps a single running count, adding 1 for every "ten" or ace coming out of the deck and subtracting 1 for every low card (2 through 6). The greater the count in the positive direction, the fewer "tens" or aces remained in the deck, favoring the player who could hit on 17 with a reduced chance of going bust. These new strategies were not only more powerful but also easier to use than Thorp's original tens strategy.

One day a senior came to my office at the beginning of his last spring term to ask me to give him a reading course in probability theory. He wanted to probe more deeply into some topics (specifically, random walks and fluctuation theory, for those who know what these terms mean) that were only touched upon in the standard courses that I taught. I should have guessed what he was up to! After a few once-a-week meetings, at which the student and I went over some fairly advanced techniques for calculating probabilities and simulating certain types of random fluctuations, I began to catch a whiff of a more than purely mathematical purpose: "Do you have any special *practical* interest in these topics?" I asked him.

With that slight prod, the student opened up and told some tales that gave me, I must admit, considerable vicarious pleasure. He and a classmate, both seniors required to take only a very light load of coursework, had been spending most of their days and nights in Las Vegas playing blackjack. They sought out single-deck games, which were still available at high-minimum tables, and they played with stacks of "quarters"—$25 chips. (My student came from a wealthy family.)

As young men playing for very high stakes, they were subject to intense scrutiny and had to go to enormous lengths to avoid being detected and barred from play. They feigned drunkenness, showed extreme interest in the cocktail waitresses (not feigned), and played with seeming lack of interest in the cards while secretly keeping their counts. They planned their assaults on the casinos with considerable care and cunning.

Every week they would pick four casinos to hit and would play blackjack for four days, sleeping on a schedule that made each day twenty hours long instead of twenty-four—a cycle that enabled them to face each eight-hour shift of casino personnel only twice in that week. The next week they would move on to another set of four casinos, taking pains never to return to the same casino until at least a month had gone by.

Being barred from play was not the only risk they faced. As Thorp's book described, some casinos were not above bringing in "cheat dealers," specialists in techniques such as "dealing seconds"—giving the player a hit with the *second* card in the deck if the top card would give him a good total. (The dealer has to peek at the top card and then execute a difficult maneuver to deal the second card instead.) Playing at a popular and very swank casino in the wee hours one morning, my student and his friend noticed that the appearance of a new dealer at the table occurred sooner than normal—a dangerous sign, according to Thorp. Suitably wary, they decided to play a few more hands and see what would happen.

My student soon faced a dealer's 10 up-card with a total of 13 in his hand, requiring him to hit. Keeping his cool in the face of possible second-dealing, he signaled for a hit. What happened next was worthy of a scene in *NUMB3RS*. The dealer moved his hand sharply to deliver the requested card, but that same motion launched another card in a high arc above the table, causing it to fall to the floor. Fortunately, the card my student was dealt was an 8, giving him a total of 21, which not surprisingly beat the dealer's total. This dramatic scene taught three lessons: that when "dealing seconds" unskilled hands might unintentionally move the top card too much to conceal the cheating; that the second card (unseen by the cheat dealer) might turn out to be even better for the player than the top card; and finally, that it was clearly time for our Caltech heroes to cash in their winnings, leave that casino, and never go back.

A few weeks after I was let in on his secret life, my student told me that he and his classmate had ended their adventures in Las Vegas. They had earned a net profit of $17,000—pretty good in those days—and they knew it was time to quit. "What makes you think so?" I asked innocently. He proceeded to explain how the "eye in the sky" system works. Video cameras are positioned above the casino ceiling to enable the casino to watch the play at the tables. They detect not only cheating but also card counting. The casino personnel who monitor the play through the camera are taught to count cards too, and by observing a player's choices, when to bet larger and smaller amounts, they can detect pretty reliably whether or not card counting is in progress.

At one well-known casino, my student and his friend returned to play after a month's absence, using all their usual techniques to avoid being spotted as card counters. They sat down at a blackjack table, bought some quarters, and placed their bets for the first hand. Suddenly, a "pit boss" (dealer supervisor) appeared, pushed their stacks of chips back to them, and politely informed them that they were no longer welcome at that casino. (Nevada law allows casinos to bar players arbitrarily.)

When my student, feigning all the innocence he could muster, asked why on earth the casino would not want to let him and his friend play a simple game of blackjack, the pit boss said, "We figure you're into us for about $700, and we're not going to let you take anymore." A full month after their last appearance, and for a mere $700. The casinos may depend on mathematics in order to make a healthy profit, but they cry foul when anyone else does the same.

APPENDIX

Mathematical Synopses of the Episodes in the First Three Seasons of *NUMB3RS*

IS THE MATH IN *NUMB3RS* REAL?

Both of us are asked this question a lot. The simplest answer is "yes." The producers and writers go to considerable lengths to make sure that any math on the show is correct, running script ideas by one or more professional mathematicians from the hundreds across the country that are listed in their address book.

A more difficult question to answer is whether the mathematics shown really could be used to solve a crime in the way depicted. In some cases the answer is a definite "yes." Some episodes are based on real cases where mathematics actually was used to solve crimes. A couple of episodes followed the course of real cases fairly closely; in others the writers exercised dramatic license with the real events in order to produce a watchable show. But even when an episode is not based on a real case, the use of mathematics depicted is generally, though not always, *believable*—it could happen. (And experience in the real world has shown that occasionally even "unbelievable" applications of mathematics do actually occur!) The skepticism critics express after viewing an episode is sometimes based on their lack of awareness of the power of mathematics and the extent to which it can be applied.

In many ways, the most accurate way to think of the series is to compare it to good science fiction: In many cases, the depiction in *NUMB3RS* of a particular use of mathematics to solve a crime is something that could, and maybe even may, happen someday in the future.

One thing that is completely unrealistic is the time frame. In a fast-paced, 41-minute episode, Charlie has to help his brother solve the case in one or two "television days." In real life, the use of mathematics in crime detection is a long and slow process. (A similar observation is equally true for the use of laboratory-based criminal forensics as depicted in television series such as the hugely popular *CSI* franchise.)

Also unrealistic is that one mathematician would be familiar with so wide a range of mathematical and scientific techniques as Charlie. He is, of course, a television superhero—but that's what makes him watchable. Observing a real mathematician in action would be no more exciting than watching a real FBI agent at work! (All that sitting in cars waiting for someone to exit a building, all those hours sifting through records or staring at computer screens . . . boring.)

It's also true that Charlie seems able to gather masses of data in a remarkably short time. In real-life applications of mathematics, getting hold of the required data, and putting it into the right form for the computer to digest, can involve weeks or months of labor-intensive effort. And often the data one would need are simply not available.

Regardless of whether a particular mathematical technique really could be used in the manner we see Charlie employ it, however, the one accurate thing that we believe comes across in practically every episode is the *approach* Charlie brings to the problems Don presents him. He boils an issue down to its essential elements, strips away what is irrelevant, looks for recognizable patterns, sees whether there is a mathematical technique that can be applied, possibly with some adaptation, or—and this has happened in several episodes—failing the possibility of applying some mathematics, at least determines whether there is a piece of mathematics that, while not applicable to the case in hand, may suggest, by analogy, how Don should proceed.

But all of the above observations miss the real point. *NUMB3RS* does not set out to teach math, or even to explain it. It's entertainment, and spectacularly successful entertainment, at that. To their credit, the writers, researchers, and producers go to significant lengths to get the math right within the framework of producing one of the most popular fictional crime series on U.S. network television. From the point of view of good television, however, it is only incidental that one of the show's lead characters is a mathematician. After all, the series is aimed at an audience that will of necessity contain a very small percentage of viewers knowledgeable about mathematics. (There are nothing like 11 million people in the country—the average *NUMB3RS* audience at an episode's first broadcast—with advanced mathematical knowledge!) In fact, Nick Falacci and Cheryl Heuton, the series' original creators and now executive producers, have observed that what persuaded the network to make and market the program in the first place was the fascination of a human interaction of two different kinds of problem solving.

Don approaches a crime scene with the street-smart logic of a seasoned cop. Charlie brings to the problem his expertise at abstract logical thinking. Bound together by a family connection (overseen by their father, Alan, played as it happens by the only family member who actually understands quite a lot of the math—Judd Hirsch was a physics major in college), Don and Charlie work together to solve crimes, giving the viewer a glimpse of how their two different approaches intertwine and interact. And make no mistake about it, the interaction of mathematical thinking with other approaches to solve problems is *very much* a real-world phenomenon. It's what has given us, and continues to give us, all of our science, technology, medicine, modern agriculture, in fact, pretty well everything we depend upon every day of our lives. *NUMB3RS* gets that right in spades.

In what follows, we provide brief, episode-by-episode synopses of the first three seasons of *NUMB3RS*. In most episodes, we see Charlie use and refer to various parts of mathematics, but in our summaries we indicate only his primary mathematical contribution to solving the case.

FIRST SEASON

1.23.05 – "Pilot"

A serial rapist/killer is loose in Los Angeles. Don leaves a map showing the crime locations on the dining table at his father's home, and Charlie happens to see it. He says he might be able to help crack the case by developing a mathematical equation that can trace back from the crime locations to identify the killer's point of origin. He explains the idea in terms of a water sprinkler, where you cannot predict where any individual droplet will land but, if you know the pattern of all the drops, you can trace back to the location of the sprinkler head. Using his equation (which you see on a blackboard in his home at one point), he is able to identify a "hot zone" where the police can carry out a sweep of DNA samples to trace the killer.

1.28.05 – "Uncertainty Principle"

Don is investigating a series of bank robberies. Charlie uses predictive analysis to accurately predict where the robbers will strike next. He likens the method to predicting the movements of fish, describing his solution as a combination of probability modeling and statistical analysis. But when Don and his team confront the thieves, a massive shootout occurs leaving four people, including an officer, dead. Charlie is devastated, and retreats into the family garage to work out a famous unsolved math problem (the P versus NP problem) that he also plunged into after his mother became terminally ill a year earlier. But Don needs his brother's help and tries to get Charlie to return to the case. When Charlie does involve himself again, he notices that the pattern of the bank robberies resembles a game called Minesweeper. The gang uses information gathered from each robbery to choose the next target.

2.4.05 – "Vector"

Various people in the L.A. area start to become sick; some of them die on the same day. Don and Charlie are called in independently (to Don's surprise) to investigate a possible bioterrorist attack, in which someone has released a deadly virus into the environment. The CDC official who calls in Charlie says they need him to help run a "vector analysis." Charlie sets out to locate the point of origin of the virus. Announcing that his approach involves "statistical analysis and graph theory," he plots all the known cases on a map of L.A., looking for clusters, and tries to trace out the infection pattern. He later explains that he is developing a "SIR model" (so-called for susceptibility, infection, recovery) of the spread of the disease, in order to try to identify "patient zero."

2.11.05 – "Structural Corruption"

Charlie believes that a college student who allegedly committed suicide by jumping from a bridge was instead murdered, and that his death is related to an engineering thesis he was working on about one of Los Angeles's newest and most important buildings, which may not be as structurally safe as the owner claims it to be. Charlie bases his suspicions on the location of the body relative to the bridge, which his calculations reveal is not consistent with the student throwing himself off the bridge. Starting with the student's data on the building, Charlie builds a computer model that demonstrates it to be structurally unsafe when subjected to certain unusual wind conditions. Suspicion falls on the foundations. By spotting numerical patterns in the company's records, Charlie determines that the records had been falsified to cover up the use of illegal immigrant workers.

2.18.05 – "Prime Suspect"

A five-year-old girl is kidnapped. Don asks for Charlie's help when he discovers that the girl's father, Ethan, is also a mathematician. When Charlie sees the mathematics Ethan has scribbled on the whiteboard in his home office, he recognizes that Ethan is working on Riemann's hypothesis, a famous math problem that has resisted attempts at solution for more than 150 years. A solution could not only earn the solver a $1 million prize, but could provide a method for breaking Internet security codes. When Don is able to determine the identity of one of the kidnappers, and learns that the plan is to "unlock the world's biggest financial secret," it becomes clear why Ethan's daughter was kidnapped.

But when Charlie finds a major error in Ethan's argument, they have to come up with a way to fool the kidnappers into believing that he really can provide the Internet encryption key they are demanding, and trace their location to rescue the daughter.

2.25.05 – "Sabotage"

A serial saboteur claims responsibility for a series of deadly train accidents. At each crash site the perpetrator leaves a numerical message, claiming in a telephone call to Don that the message tells him everything he needs to know about the series of crashes. The FBI team assumes the message is in a numeric code, which Charlie tries to crack. Charlie sees lots of numerical patterns in the message but is unable to crack the code. Charlie and the FBI team soon realize that each accident was a re-creation of a previous wreck, and eventually Charlie figures out that there is no code. The message is a compendium of data about a previous crash. Charlie says, "It's not a code, it's a story told in numbers."

3.11.05 – "Counterfeit Reality"

A team of forgers has taken an artist hostage to draw the images to produce small-denomination counterfeit bills. The counterfeiters murder at least five people, leading Don to believe that if the missing artist isn't located soon she will be killed when she finishes her work on the phony money. Charlie is brought in to run an algorithm to enhance the image quality on some store-security videotapes relevant to the case. After studying the fake bills, he notices some flaws that appear to be deliberate, but do not seem to have any pattern. His student Amita suggests that if he looks at the image at an angle, he may be able to discern a pattern. In this way, he is able to read a secret clue, written by the kidnapped artist, that leads the FBI to the gang's location.

4.1.05 – "Identity Crisis"

A man wanted for stock fraud is found garroted in his apartment, and the crime is eerily similar to a murder committed a year earlier, a case which Don closed when an ex-con confessed. Now, Don must re-investigate the old case to determine whether he put an innocent man in jail. He asks Charlie to go over the evidence to see if he missed anything the first time around. Charlie questions the procedure used for identification of suspects from photographs and the method of using fingerprints for identification. He carries out a statistical analysis of eyewitness evidence reliability.

4.15.05 – "Sniper Zero"

Los Angeles is plagued by a spate of sniper killings. Charlie initially tries to determine the location of the sniper by calculating the trajectories of the bullets found in the victims, mentioning his use of "drag coefficient models." By graphing the data and selecting axes appropriately, Charlie concludes that more than one shooter is at work. He suspects that the data is following an exponential curve, suggesting that there is an epidemic of sniper attacks, inspired by an original "sniper zero." He compares the situation to the decisions of homeowners to paint their houses a certain color, mentioning the much discussed "tipping point" phenomenon. He analyzes the accuracy of the shooters in terms of "regression to the mean," and concludes that the key pattern of sniper zero is not in the locations of the victims but in where the sniper fired the shots.

4.22.05 – "Dirty Bomb"

A truck carrying radioactive material is stolen, and the thieves threaten to set off a dirty bomb in L.A. in twelve hours if they aren't paid $20 million. While Don attempts to track down the truck, Charlie analyzes possible radiation dispersal patterns to come up with the most likely location where the bomb may be detonated to inflict the most damage to the population. However, the gang's real aim is for the FBI to evacuate an entire city square, in order to steal valuable art from a restoration facility. Eventually the FBI is able to identify and capture the three criminals, who use the threat of detonating a dirty bomb to try to negotiate their release. Observing that the isolation and individual interrogation of the three criminals is reminiscent of the so-called prisoner's dilemma, Charlie has the three brought together to present them with a risk-assessment calculation, which shows how much each has to lose. This causes the one with the greatest potential loss to come clean and say where the radioactive material is hidden.

4.29.05 – "Sacrifice"

A senior computer-science researcher, working on a classified government project, is found murdered in his Hollywood Hills home. The FBI discovers that data had been erased from the dead man's computer around the time of the murder. Don's investigation reveals that the victim was going through a bitter divorce, and was trying to keep his wife from getting his money. Using what he refers to as a predictive equation, Charlie is able to recover enough data from the victim's erased hard drive to learn that the project the man was working on seemed to involve baseball statistics. But when Charlie runs a Google search on some of the number sequences, he discovers that the data came not from baseball but from government statistics on people living in different kinds of neighborhoods.

5.6.05 – "Noisy Edge"

Together with an agent from the National Transportation Security Board, Don investigates eyewitness accounts of a mysterious unidentified object flying dangerously close to downtown Los Angeles, that has raised concern of a terrorist attack. After Charlie is recruited to help with the investigation, they discover that the flying object is part of a new technology that could revolutionize air travel. But the investigation takes a more sinister turn when they discover evidence suggesting sabotage that leads to the crash of the aircraft, killing the lead engineer who was piloting the plane on a test flight. There is considerable discussion of the "squish-squash algorithm," developed by a mathematician at the University of Alberta to uncover weak signals (such as radar) in a noisy environment.

5.13.05 – "Manhunt"

As Don investigates a prison bus crash, Charlie uses probability analysis to conclude that the bus crash was not an accident, but part of a conspiracy to free a dangerous killer who is bent on revenge. Don and Charlie must find the killer before he is able to carry out his intent. Charlie uses probability theory to try to predict where the killer is likely to go next. This involves the use of Bayesian analysis to determine which of the many reported sightings of the fugitive by the public are more likely to be reliable. He uses the results to plot places and times to furnish a trajectory.

SECOND SEASON

9.23.05 – "Judgment Call"

The wife of a federal judge is shot and killed in her garage. It's unclear whether the intended target was her or her husband, who was hearing a death penalty case involving a gang leader. Don wants to know which of the many criminals the judge has sent to prison are most likely to seek revenge. Charlie's task is to narrow down the list of possible suspects. He initially refers to his approach as using a "Bayesian filter" and later talks about "reverse decision theory," Presumably what he is doing is using Bayes' theorem "backwards," to compute for each suspect the probability that he or she committed the murder, so that Don can concentrate on the ones to whom Charlie's calculations assign the highest probabilities.

9.30.05 – "Better or Worse"

A young woman attempts to rob a jewelry store in Beverly Hills by showing the store owner a photograph of his kidnapped wife and child. As the woman is leaving the store with a large quantity of diamonds, she is shot and killed by a security guard. Charlie assists the FBI by cracking the code of the keyless remote from the woman's car, found in her purse, to help identify her through her car purchase, and hence locate and rescue the store owner's kidnapped wife and daughter. Since the security of car remotes depends on sequences of numbers, the "obvious" mathematical approach is to look for numerical patterns that provide a clue to the entire code. Presumably this is what Charlie does, but he never specifies the techniques he is using.

10.7.05 – "Obsession"

The young wife of a high-profile Hollywood movie producer is stalked while alone in her home. The house is fitted with an extensive system of security cameras, but none of them has recorded any image of the intruder. Charlie realizes that the intruder must know the house and the location of the cameras, and is using a laser to temporarily "blind" the cameras as he passes in front of them. This leads him to analyze the video recordings using sophisticated image enhancement algorithms that are able to generate a reliable image of the stalker from relatively little information.

10.14.05 – "Calculated Risk"

Clearly inspired by the Enron case. A whistle-blower is killed, the financial officer of a large energy company who had exposed a major financial fraud. The problem facing Don is the sheer number of people with a motive to kill her: the senior people at the company who want to prevent her from testifying against them in court, the thousands of company employees who will lose their jobs if the company goes under, and the still greater number of people who are likely to lose most of their pension. Charlie uses a technique called "tree pruning" to narrow down a probabilistic suspect relationship tree from all of those affected by the swindle. He then models the flow of money through the company using methods of fluid flow in order to identify the killer.

10.21.05 – "Assassin"

During an arrest of a forger, Don uncovers a notebook containing encoded entries. He asks Charlie if he can decipher the contents.

Drawing on his background of consulting for the NSA, Charlie is able to crack the code, and discovers that the notebook contains plans for a skilled and trained assassin to murder a Colombian exile living in Los Angeles. His remaining contribution to the case is to suggest to Don ways to pursue the assassin based on ideas from game theory, speculating on how the killer will behave in different situations.

11.4.05 – "Soft Target"

A Homeland Security exercise in the Los Angeles Metro turns into a real emergency when someone releases phosgene gas in a train. Don is assigned to the case. Using classical percolation theory (based on statistical mechanics, which determines the flow of liquids and gases based on the motion of the individual molecules) to determine the flow of the gas, based on the readings from the people in the car, Charlie figures out the precise location where it was released. After Don identifies a likely suspect, Charlie tries to predict where and how he will strike next, by applying linear percolation theory, a fairly new field which Charlie explains in terms of a ball running through a pinball machine.

11.11.05 – "Convergence"

A chain of robberies at upscale Los Angeles homes takes a more sinister turn when one of the homeowners is murdered. The robbers seem to have a considerable amount of inside information about the valuable items in the houses robbed and the detailed movements of their owners. Yet the target homes seem to have nothing in common, and certainly nothing that could provide a source to the detailed information the crooks are clearly getting. Charlie approaches the task using data-mining techniques, applying data-mining software to look for patterns among all robberies in the area over the six-month period of the home burglaries. Eventually he comes up with a series of car thefts that look as though they could be the work of the same gang, and this leads to their capture. His other contribution to the case is figuring out that the gang keeps track of the homeowners' movements by intercepting signals from the GPS location chip found in all modern cell phones.

11.18.05 – "In Plain Sight"

A raid on a methamphetamine lab goes wrong and an FBI agent is killed when the booby-trapped house blows up. The lab was identified in part by Charlie's analysis of social networks using flocking algorithms. Attempts to enhance a photographic image from a computer found at the house reveal a child pornography image encoded using steganography. Further analysis of the computer hard drive yields a hidden partition, the contents of which provide a clue to the leader of the meth lab.

11.25.05 – "Toxin"

An unknown person is spiking certain over-the-counter medications with poisons. This soon leads Don and his team to a hunt for a fugitive who has disappeared into the California mountains. Charlie takes inspiration from information theory and from combinatorics (Steiner trees) to help Don solve the case. The mathematics is not so much applied as used to provide an illustration of what actions Don should take.

12.9.05 – "Bones of Contention"

The discovery of an ancient skull leads to the murder of a museum antiquarian. Charlie uses his knowledge of carbon dating and Voronoi diagrams (a concept in combinatorics related to the efficient distribution of goods) to help solve the crime. The carbon-dating part is a now standard application of mathematics to determine the age of death associated with skeletons and bone fragments. The Voronoi diagram part is not unlike Charlie's mention of Steiner trees in the previous episode, "Toxin": it is more a way of focusing attention on a key aspect of the investigation.

12.16.05 – "Scorched"

An arsonist sets a fire at an SUV dealership that kills a sales person. The name of an extremist environmental group is spray-painted on the scene, but the group denies involvement. Don has to determine whether the group is responsible or someone else set the fire. Charlie is called in to help figure out whether there is a pattern to the fires that would help provide a profile of the arsonist. He says he is using "principal components analysis" to produce arson "prints" that will be sufficiently precise to identify the criminal.

1.6.06 – "The O.G."

An FBI agent working undercover as a gang member is killed. When it appears that his cover had not been blown, it begins to look like yet another round in an ongoing battle between rival gangs. Charlie thinks that with so many gang killings, 8,000 over four years, there is enough data to use social network analysis to look for tit-for-tat chains of killings. His analysis uncovers several chains much longer than the average chains, and he thinks they are likely to be the work of the same killer or killers. His detection of unusual features of some of the chains eventually enables Don to solve the case. The episode title stands for the term "old gangster."

1.13.06 – "Double Down"

When a young man who is killed just after leaving a casino with considerable winnings turns out to be a brilliant mathematics student at a local university, Don suspects that the victim was part of a group of players using "card counting" to improve their chances of winning. Charlie's analysis takes into account the latest developments in the fifty-year history of using mathematical analyses to win at blackjack.

1.27.06 – "Harvest"

A report of suspicious activity in the basement of a hotel leads Don to uncover a black market scheme trading in body parts. Young girls from a poor area of rural India are persuaded to sell, say, a kidney, to be transplanted to a wealthy patient in Los Angeles. The girls are brought over, the operation performed, and then they are sent back. But after one of the girls dies, Don worries that the gang will feel they have nothing more to lose if others die, too. Charlie's contribution is to determine the most likely time of the girl's death based on photographs of a pile of partially melted ice taken by the police when they arrived on the scene. The ice would have been brought in to preserve the kidney in transit, and would have been fresh at the time of the operation on the girl.

2.3.06 – "The Running Man"

A gang steals a DNA synthesizer from CalSci, and Don suspects that the thieves intend to sell it to a terrorist group that would use it to manufacture biological weapons. Charlie provides assistance (in a very minor way) by suggesting a possible analogy with Benford's Law, which describes a surprising distribution of leading digits in tables of real-world data (1 thirty percent of the time, 2 eighteen percent of the time, 3 twelve percent, down to 9 a mere four percent). Naïve intuition would suggest that with randomly distributed figures, each digit would occur one-ninth of the time, but this is not so for data from a real-world source. In the case Don is working on, the equivalent of the prevalent leading digit turns out to be CalSci's LIGO lab, which Larry directs. LIGO stands for "Laser Interferometer Gravitational-Wave Observatory". (Caltech—the real-world "CalSci"—actually does operate a LIGO lab, though the facility itself is not located on their campus, or even in California.)

3.3.06 – "Protest"

Don and his team investigate an antiwar bombing outside an Army recruiting center that resembles the work of a 1970s antiwar activist who, thirty-five years earlier to the day, had planted a bomb that killed two people. That bomber was never caught and the FBI's principal suspect at the time had disappeared soon after the explosion. Charlie uses social network analysis to help Don figure out who might have carried out the 1971 bombing, leading to an unexpected discovery about the undercover activities of the FBI in the anti–Vietnam War movement.

3.10.06 – "Mind Games"

Following leads provided by a self-proclaimed psychic, a search team finds three dead girls in the wilderness. The victims, all illegal immigrants, were apparently murdered under bizarre, ritualistic circumstances, but it is later revealed that they were killed to recover illegal drugs they had smuggled across the Mexican border inside their stomachs. Much of Charlie's activity in the episode is devoted to trying to persuade Don and the others that there is no such thing as ESP and that people who claim they are psychics are frauds. He does however contribute to the solution of the case by using the Fokker-Planck equation (which describes the chaotic motion of a body subject to certain forces and constraints) to determine where the next group of smugglers may be hiding out.

3.31.06 – "All's Fair"

An Iraqi woman, a human rights activist in Los Angeles to make a documentary promoting the rights of Muslim women, is murdered. Charlie examines the statistical records of many possible suspects to try to find the ones most likely to have committed the crime. To do this, he has to weigh all the factors that might indicate a willingness to murder. This enables him to give each suspect a "score" or probability, with the ones having the highest scores the primary suspects. Creating a weighting in this fashion, based on statistics, is called statistical regression, and the particular type that Charlie uses is called "logistic" regression.

4.7.06 – "Dark Matter"

Don and his team investigate a high school shooting in which eight students were killed, along with one of the shooters. The school has a radio frequency identification system to track the movements of each pupil throughout the day, and Charlie uses the recorded data from the system to track the movements of the shooters and their victims through the school's hallways, using "predator-prey" equations. When his analysis uncovers an abnormal pattern, Charlie is sure there was a third shooter that no one had suspected earlier.

4.21.06 – "Guns and Roses"

A government law enforcement agent is found dead in her home. At first it looks like suicide, but when details of the woman's recent investigation and private life start to emerge, Don grows suspicious. Charlie uses acoustic fingerprinting, based on recordings of the gunshot picked up by police radios in the area, and concludes that there must have been another person in the room at the time the agent died. Acoustic fingerprinting has been used on several occasions in actual shootings, including the 1963 Kennedy assassination, where the mathematical analysis indicated the high probability of a second shooter firing from the famous "grassy knoll."

4.28.06 – "Rampage"

A man steals a gun from an agent in the FBI office and starts shooting people at random. After agent David Sinclair overpowers him, it is discovered that he is a respectable husband and father, seemingly without motive. After considerable investigation, Don learns that the man was a pawn in an elaborate scheme to derail an upcoming trial of a dangerous arms dealer. Charlie provides a key step in the investigation by determining how closely the shooter's path resembled Brownian (random) motion. He also uses an analogy with a four-dimensional hypercube to motivate an examination of the shooting as a spacetime event.

5.5.06 – "Backscatter"

Don investigates a computer hacking scam that breaks into a bank's system to gain access to the identities and financial assets of its customers, including Don. It turns out that the Russian mafia is behind the activity. Interestingly, although the security of bank computer and data systems depends on masses of advanced mathematics, some of which Charlie mentions, the solution to this case does not make much use of math—it's all "beneath the hood," buried in the tracking systems that Charlie and Amita use to help Don.

5.12.06 – "Undercurrents"

The bodies of several young Asian girls are washed up on the beach, possibly having been thrown overboard. The situation becomes more critical when it is discovered that one girl has avian flu. Charlie carries out some calculations focused on ocean currents to determine the most likely location where the victims entered the water. As the investigation continues, Don and his team discover a connection between the girls and the sex trade industry.

5.19.06 – "Hot Shot"

Don investigates the murders of two young women, found in their cars outside their homes. Their deaths were made to look like drug overdoses, but Don soon concludes that a serial killer is responsible. Charlie tries to help by analyzing the daily routines of the two women, looking for patterns that might provide leads to the killer, but Don solves the case largely by standard investigative techniques.

THIRD SEASON

9.22.06 – "Spree"

The first installment of a two-part season-opener. A young couple embark on a cross-country spree of robberies and murders. When it becomes clear that their movements are influenced by the pursuit of an FBI agent, who joins forces with Don and his team, Charlie uses "pursuit curves" to help the agents track them down. The effectiveness of the mathematics becomes critical after one of the fugitives is caught and the remaining one kidnaps Agent Reeves to trade her for her partner.

9.29.06 – "Two Daughters"

This is the completion of the previous week's episode, "Spree."

10.6.06 – "Provenance"

A thief steals a valuable painting from a small local art gallery. The case turns more sinister when one of the key suspects is murdered. Charlie analyzes a high-resolution photograph of the missing painting through the use of mathematical techniques and, by comparing his results with a similar analysis of other paintings by the same artist, he concludes that the stolen painting is a fake, leading Don to revise his suspect list. His analysis uses a method developed by a (real) mathematician at Dartmouth College, which reduces the fine details of the painting (relative areas of light and dark, choice of colors, perspective and shapes used, width, thickness and direction of brush strokes, shapes and ridges within brushstrokes, etc.) to a series of numbers—a numerical "fingerprint" of the painter's technique.

10.13.06 – "The Mole"

An interpreter at the Chinese consulate is killed in a hit-and-run traffic accident. When Charlie carries out a mathematical analysis on how she must have been hit, it is clear she was murdered. When Don investigates the dead girl, he discovers that she was probably working as a spy. Though Charlie also provides assistance by using the facial recognition algorithm he has been developing, as well as using steganography extraction algorithms to reveal messages hidden in computer images, Don and his team solve the case largely without Charlie's involvement, using more traditional, nonmathematical techniques.

10.20.06 – "Traffic"

Don investigates a series of attacks on L.A. highways. Are they coincidence or the work of a single attacker? Are some of them copycat attacks? Charlie and Amita first help by analyzing traffic flow using the mathematics of fluid flow, a technique frequently used in real-life traffic-flow studies. But Charlie's main contribution comes when it is suggested that the characteristics of the attacks and the choices of victims seem too random. He examines the pattern of the crimes and convinces Don that they must be the work of a single perpetrator. The challenge then is to find the hidden common factor that connects the victims.

10.27.06 – "Longshot"

This was one of the rare *NUMB3RS* episodes where they got the math badly wrong. A young horse-race gambler is murdered at the racetrack. It turns out that the bettor had made thirty bets on thirty races over the past five days and won them all. This is such an unlikely occurrence mathematically that all the races must have been rigged, yet Charlie, who is usually right on the mathematical ball, never makes that observation. If he had, Don, always on the real-life-knowledge-of-the-world ball, would doubtless have gone on to say that there is no way even organized crime could rig so many races. All in all, from a mathematical perspective and in terms of believability, this episode misfired. 'Nuff said.

11.3.06 – "Blackout"

A series of failures at electricity substations cause localized blackouts in areas of Los Angeles. Don worries that a terrorist group is running trials prior to launching an attack intended to cause a cascading failure that will plunge the entire city into darkness. But when Charlie analyzes the flow network, he discovers that none of the targets would have such an effect, and suspects that the attacks have a different purpose. By analyzing the target substations and the ones left alone using elementary set theory (Venn diagrams and Boolean combinations), he is able to identify the real target—a prison housing a man waiting for trial, whom various other criminals would prefer to see dead.

11.10.06 – "Hardball"

The sudden death of an aging baseball player during practice turns sinister when steroids found in his locker turn out to be in a lethal dosage that can only be deliberate. The player was murdered. The discovery of what seem to be ransom-threat e-mails to the murdered player brings Charlie into the picture because the unknown e-mailer based his accusations on a mathematical analysis of the player's performance that indicated exactly when he started using steroids. Initial suspicion falls on a young baseball fan who uses sabermetrics (the mathematical analysis of baseball performance statistics) to play fantasy baseball. The key mathematical idea that led the young fan to spot the steroid use is called changepoint detection.

11.17.06 – "Waste Not"

When a sinkhole opens in a school playground, killing one adult and injuring several children, Don is called in because the company that constructed the playground had been under investigation for suspected negligence. Charlie's analysis of health issues in the Los Angeles region turns up unusually high incidences of childhood cancers and other illnesses concentrated in areas where the company had constructed a playground using an asphalt substitute made from recycled toxic waste. The material seems harmless, but when Charlie spots a discrepancy between the waste material sent to the company and the surfacing material produced, he suspects that drums of untreated waste had been buried beneath the playgrounds. Charlie uses reflection seismology to locate some of the buried drums. This is a method for obtaining an image of the terrain beneath the surface by mathematically analyzing the reflections of shock waves from a small underground explosion.

11.24.06 – "Brutus"

A California state senator and a psychiatrist are murdered. The two cases appear very different, but Don thinks the two murders are related. Charlie helps by using network theory to unearth possible connections between the two victims. The trail leads to a long-kept government secret. The episode opens with Charlie testing a crowd-monitoring surveillance system he has developed, based on the mathematics of fluid flow.

12.15.06 – "Killer Chat"

Charlie helps Don track a killer who has murdered several sexual predators. The predators had all taken advantage of teenage girls they met in online chat rooms, and the killer lured them to their deaths by posing online as a teenage girl. Charlie's principal contribution is to analyze the linguistic patterns of the various participants in the chat, captured by the chat-room logs, a technique often used in real-life law enforcement.

1.5.07 – "Nine Wives"

Don, Charlie, and the team search for a polygamist who is on the run. The man is on the FBI's "Ten Most Wanted" list for rape and murder. The events of this episode closely mirror those of the real-life case of Warren Steed Jeffs, and the fictitious cult "Nine Wives" was based upon the Fundamentalist Church of Jesus Christ of Latter Day Saints (FLDS), of which Warren Steed Jeffs was the leader. Charlie's principal contribution comes when he analyzes a network diagram found at one of the cult's hideouts, which his department chair, Millie, recognizes as a genetic descendant graph.

1.12.07 – "Finders Keepers"

When an expensive, high-performance racing yacht sinks during the middle of a race, Don is not the only one who gets involved. Agents from the NSA show up on the scene as well. Charlie helps by using fluid dynamics equations to calculate the most likely location where the vessel can be found. When it eventually turns up somewhere else, it becomes clear that there is far more to the story than first appeared. Charlie carries out a further analysis of the yacht's journey and concludes that it must have been carrying a heavy cargo hidden in the keel. The NSA agents are forced to disclose what brought them into the picture.

2.2.07 – "Take Out"

A gang has been robbing patrons of upscale restaurants, killing diners in the process. Charlie analyzes the pattern of locations of the restaurants to try to figure out where they are most likely to strike next. When the gang strikes at another restaurant, not on Charlie's list, he has to reexamine his assumptions. It soon becomes clear that there is more to the robberies than simply the acquisition of money. To track them down, Charlie has to find a way to trace the flow of funds through off-shore banks that serve as money-laundering operations.

2.9.07 – "End of Watch"

Don and his team reopen a cold case when an LAPD badge turns up at a construction site. Charlie uses a highly sophisticated (and math heavy) technique called "laser swath mapping" to locate the buried remains of the owner of the badge, an officer who has been missing seventeen years. LSM uses a highly focused laser beam from low-flying aircraft to identify undulations in the ground. Later in the episode, Charlie uses critical path analysis to try to reconstruct the dead officer's activities on the day he died. The episode title, "End of Watch," is a police idiom for the death of a cop. At police funerals, "end of watch" is used to indicate the date that an officer passed away.

2.16.07 – "Contenders"

One of David's old school friends kills a sparring partner in the ring. It looks like an accident until it emerges that the same thing has happened before. When the coroner discovers that the dead fighter was poisoned, things look bad for David's friend, but a DNA analysis of some key evidence eventually clears him. Charlie says he can use a "modified Kruskal count" to analyze the sequence of fights the two dead fighters were involved in to determine the likely killer. A Kruskal count is a device for keeping track of playing cards used by stage magicians to "predict" the face value of a card that has apparently been lost in a sequence of shuffles. It is hard to see how this technique could be used in the way Charlie suggests. Perhaps his mind was distracted by the upcoming poker championship he is playing in, which provides a secondary theme for this episode.

2.23.07 – "One Hour"

While Don is occupied talking with the agency's psychiatrist, his team is in a race against time to find an eleven-year-old boy, the son of a wealthy local gangster who has been kidnapped for a $3 million ransom. Much of the action centers on the kidnapper directing agent Colby Granger to follow a complex path through Los Angeles to shake off any tails, a sequence taken from the Clint Eastwood movie *Dirty Harry*. Charlie and Amita assist by figuring out the logic behind the route the kidnapper has Colby follow, though it is never made clear quite how they manage this. It seems unlikely given the relatively small number of data points.

3.9.07 – "Democracy"

Several murders in the Los Angeles area seem to be tied to election fraud using electronic voting machines. Don, Charlie, and the team must find the killers before they strike again. Although the security of electronic voting systems involves lots of advanced mathematics, Charlie's principal contribution to solving this case is right at the start, when he computes the likelihood that a particular sequence of deaths could be accidental. When his answer turns out to be extremely low, that provides Don the key information that the deaths were all murders.

3.30.07 – "Pandora's Box"

A small executive jet crashes in the forest, witnessed by a forest ranger. When the ranger goes to investigate, he is shot, raising the suspicion of sabotage. The black box recorder is recovered and analyzed (by Charlie in a CalSci lab) and shows that the plane's altitude readings were off by several thousand feet. By analyzing the debris field, Charlie is able to locate the aircraft's flight control computer. When he analyzes the code, he discovers that the entire crash was caused as a ruse to insert computer code into the FAA's main flight control computer when the black box was read. Charlie's other main contribution to solving the case is the use of image enhancement techniques to deblur some key smudged fingerprints.

4.6.07 – "Burn Rate"

A series of letter bombs protesting biotechnology research has the same features of an earlier series for which someone is already serving a prison sentence. Charlie's initial contribution is to analyze the debris from the explosions to determine the construction of the bomb. Then he looks at the pattern of addresses from which the bombs were mailed to narrow down the main suspect's likely location. But when he realizes the data is simply too good—there are no outliers—he realizes that Don's suspect cannot be the bomber. But who is?

4.27.07 – "The Art of Reckoning"

A former mob hit man on death row has a change of heart and agrees to confess to his crimes in exchange for seeing his daughter before he is executed. Charlie advises Don how to conduct the negotiation by explaining the tit-for-tat strategy for repeated plays of the Prisoner's Dilemma, a two-person competitive game. The use of an fMRI scanner to determine if the condemned man is lying depends on a lot of sophisticated mathematics, but it's all buried in the technology, so Charlie does not have to do it.

5.4.07 – "Under Pressure"

Information recovered from a laptop obtained in Yemen indicates that a team of terrorists intends to pump nerve gas into the Los Angeles water supply. Charlie uses network analysis to try to figure out who the key operatives might be. Most of Charlie's contribution has occurred before the episode starts.

5.11.07 – "Money for Nothing"

A truck carrying medicines and fifty million dollars in cash destined for an African relief program is hijacked by a gang of sophisticated thieves. The FBI's efforts to locate the shipment are complicated by the activities of bounty hunters. Charlie performs a mathematical analysis of the truck's possible escape paths.

5.18.07 – "The Janus List"

A former cryptologist for British intelligence agencies confronts the FBI and sets off fiery explosions on a bridge as part of a desperate scheme to expose double agents who have poisoned him. To help the FBI follow the cryptologist's complicated trail of clues and make the critical contacts needed to obtain the list of double agents, Charlie must decipher messages that have been encoded using a variety of techniques, including a straddling checkerboard and a musical cipher.

Index

Credits

Page 133 courtesy of the authors

Page 141 courtesy of Valdis Krebs, www.orgnet.com

Page 144 courtesy of Gary Lorden

Page 165 courtesy of Lawrence M. Wein of Stanford University